Syntactic Methods
in Pattern Recognition

This is Volume 112 in
MATHEMATICS IN SCIENCE AND ENGINEERING
A series of monographs and textbooks
Edited by RICHARD BELLMAN, *University of Southern California*

The complete listing of books in this series is available from the Publisher
upon request.

Syntactic Methods in Pattern Recognition

K. S. FU

School of Electrical Engineering
Purdue University
West Lafayette, Indiana

 *1974*

ACADEMIC PRESS   New York   San Francisco   London

A Subsidiary of Harcourt Brace Jovanovich, Publishers

ACADEMIC PRESS, INC.
111 Fifth Avenue, New York, New York 10003

United Kingdom Edition published by
ACADEMIC PRESS, INC. (LONDON) LTD.
24/28 Oval Road, London NW1

Library of Congress Cataloging in Publication Data

Fu, King Sun, Date
    Syntactic methods in pattern recognition.

    (Mathematics in science and engineering, v. 112)
    Includes bibliographical references.
    1.   Pattern perception.   2.   Mathematical linguistics.
I.   Title.   II.   Series.
Q327.F82   001.53'4   73-18948
ISBN 0-12-269560-7

Contents

Chapter 3 Languages for Pattern Description

Chapter 4 Syntax Analysis as a Recognition Procedure

Chapter 5 Stochastic Languages and Stochastic Syntax Analysis

Preface

During the past fifteen years there has been a considerable growth of interest in problems of pattern recognition. This interest has created an increasing need for methods and techniques for use in the design of pattern recognition systems. Contributions to the blossoming of this area have come from many disciplines, including statistics, psychology, linguistics, computer science, biology, taxonomy, switching theory, communication theory, control theory, and operations research. Many different approaches have been proposed and a number of books have been published. Most books published so far deal with the so-called decision-theoretic or discriminant approach. Recently, perhaps stimulated by the problems of picture recognition and scene analysis, the syntactic or structural approach has been proposed, and some preliminary results from applying this approach have shown it to be quite promising. This monograph treats the problems of pattern recognition by use of the syntactic approach. It is intended to be of use both as a reference for systems engineers and computer scientists and as a supplementary textbook for courses in pattern recognition. The presentation is kept concise, and background information on formal languages and syntax analysis is included.

With the emphasis on structural description of patterns, the syntactic approach attempts to draw an analogy between the structure of patterns and the syntax of a language. The analogy is attractive primarily due to the availability of mathematical linguistics as a tool. It is true that linguistic methods are syntactic in nature. However, the syntactic approach certainly contains other nonlinguistic methods, though the materials discussed in this monograph are mainly linguistic. From the viewpoint of pattern description or modeling, class distribution or density functions are used to describe patterns in each class in the decision-theoretic approach, but syntactic rules or grammars are employed to describe patterns in the syntactic approach. The effectiveness of these approaches appears to be dependent upon the particular problem at hand. Often a mixed approach needs to be applied. As a matter of fact, it is sometimes difficult to distinguish sharply between syntactic and nonsyntactic pattern recognizers.

The subject matter may be divided into three major parts: (1) syntactic pattern recognition using formal languages and their higher dimensional generalizations, (2) syntactic pattern recognition using stochastic languages, and (3) learning in syntactic pattern recognition—grammatical inference. In Chapter 1 the syntactic approach to pattern recognition is introduced and several pattern preprocessing and segmentation techniques are briefly reviewed. An introduction to formal languages and their relations to automata (acceptors) are summarized in Chapter 2. Various syntactic techniques for pattern description that include the use of string languages and their higher dimensional extensions are presented in Chapter 3. Syntax analysis and its applications as pattern recognition procedures are discussed in Chapter 4. Chapter 5 provides a review of existing results in stochastic languages and stochastic syntax analysis. Examples of applying stochastic languages to syntactic pattern recognition are given in Chapter 6. An introduction and survey of grammatical inference and some preliminary applications to problems in syntactic pattern recognition are presented in Chapter 7. Appendixes A to I provide a number of examples to demonstrate the applications of the syntactic approach.

Acknowledgments

It is the author's pleasure to acknowledge the encouragement of Dr. M. E. Van Valkenburg, Dr. L. A. Zadeh, Dr. E. S. Kuh, Dr. J. C. Hancock, Dr. C. L. Coates, Jr., and Mr. E. Schutzman. Some of the material in the monograph has been taught at Purdue University, the University of California at Berkeley, and Stanford University. The author is indebted to his colleagues and students at Purdue University, the University of California at Berkeley, and Stanford University, who, through many helpful discussions during office and class hours, lunch and coffee breaks, and late evenings, assisted in the preparation of the manuscript. Discussions with A. Rosenfeld, R. H. Anderson, and T. Pavlidis on syntactic pattern recognition have been always rewarding and fruitful. The author has been very fortunate in having consistent support from the National Science Foundation and the Air Force Office of Scientific Research for research in patternrecognit ion. The major part of the manuscript was completed during the author's sabbatical year (1972) at Stanford University and the University of California at Berkeley on a Guggenheim Fellowship. The author owes a debt of gratitude for the support of the John Simon Guggenheim Memorial Foundation. In addition, the author wishes to thank Ms. Kathy Mapes and Ms. Terry Brown for their efficient and careful typing of the manuscript.

Chapter 1

Introduction

1.1 Syntactic (Structural) Approach to Pattern Recognition

The many different mathematical techniques used to solve pattern recognition problems may be grouped into two general approaches; namely, the decision-theoretic (or discriminant) approach and the syntactic (or structural) approach. In the decision-theoretic approach, a set of characteristic measurements, called features, are extracted from the patterns; the recognition of each pattern (assignment to a pattern class) is usually made by partitioning the feature space [1]. Most of the developments in pattern recognition research during the past decade deal with the decision-theoretic approach [1–11] and its applications. In some pattern recognition problems, the structural information which describes each pattern is important, and the recognition process includes not only the capability of assigning the pattern to a particular class (to classify it), but also the capacity to describe aspects of the pattern that make it ineligible for assignment to another class. A typical example of this class of recognition problem is picture recognition or, more generally speaking, scene analysis. In this class of recognition problem, the patterns under consideration are usually quite complex and the number of features required is often very large, which makes the idea of describing a complex pattern in terms of a (hierarchical) composition of simpler subpatterns very attractive. Also, when the patterns are complex and the number of possible descriptions

1

is very large, it is impractical to regard each description as defining a class (e.g., in fingerprint and face identification problems; recognition of continuous speech or of Chinese characters). Consequently, the requirement of recognition can only be satisfied by a description for each pattern rather than by the simple task of classification.

Example 1.1 The pictorial patterns shown in Fig. 1.1 can be described in terms of the hierarchical structures shown in Fig. 1.2.

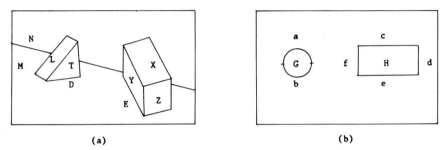

(a)   (b)

Fig. 1.1. The pictorial patterns for Example 1.1. (a) Scene *A*; (b) Picture *F*.

In order to represent the hierarchical (treelike) structural information of each pattern, that is, a pattern described in terms of simpler subpatterns and each simpler subpattern again described in terms of even simpler subpatterns, etc., the syntactic or structural approach has been proposed [11–16]. This approach draws an analogy between the (hierarchical, or treelike) structure of patterns and the syntax of languages. Patterns are specified as being built up out of subpatterns in various ways of composition, just as phrases and sentences are built up by concatenating words, and words are built up by concatenating characters. Evidently, for this approach to be advantageous, the simplest subpatterns selected, called "pattern primitives," should be much

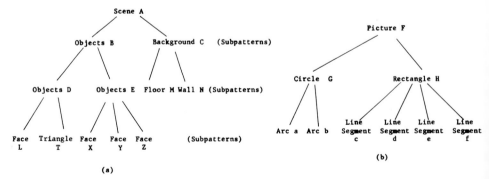

Fig. 1.2. Hierarchical structural descriptions of (a) Scene *A* and (b) Picture *F*.

easier to recognize than the patterns themselves. The "language" that provides the structural description of patterns in terms of a set of pattern primitives and their composition operations is sometimes called the "pattern description language." The rules governing the composition of primitives into patterns are usually specified by the so-called grammar of the pattern description language. After each primitive within the pattern is identified, the recognition process is accomplished by performing a syntax analysis, or parsing, of the "sentence" describing the given pattern to determine whether or not it is syntactically (or grammatically) correct with respect to the specified grammar. In the meantime, the syntax analysis also produces a structural description of the sentence representing the given pattern (usually in the form of a tree structure).

The syntactic approach to pattern recognition provides a capability for describing a large set of complex patterns by using small sets of simple pattern primitives and of grammatical rules. As can be seen later, one of the most attractive aspects of this capability is the use of the recursive nature of a grammar. A grammar (rewriting) rule can be applied any number of times, so it is possible to express in a very compact way some basic structural characteristics of an infinite set of sentences. Of course, the practical utility of such an approach depends on our ability to recognize the simple pattern primitives and their relationships, represented by the composition operations.

The various relations or composition operations defined-among subpatterns can usually be expressed in terms of logical and/or mathematical operations. For example, if we choose "concatenation" as the only relation (composition operation) used in describing patterns, then for the pattern primitives shown in Fig. 1.3a the rectangle in Fig. 1.3b would be represented by the string *aaabbcccdd*. More explicitly, if we use $+$ for the "head-to-tail concatenation"

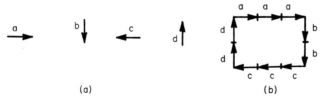

(a)   (b)

Fig. 1.3. A rectangle and its pattern primitives.

operation, the rectangle in Fig. 1.3b would be represented by $a + a + a + b + b + c + c + c + d + d$, and its corresponding treelike structure would be that shown in Fig. 1.4a. Similarly, a slightly more complex example is given in Fig. 1.4b, using the pattern primitives in Fig. 1.3a.

An alternative representation of the structural information of a pattern is a "relational graph." For example, a relational graph of Picture F in Fig. 1.1b

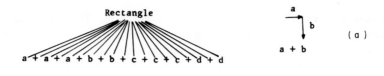

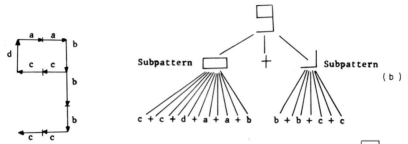

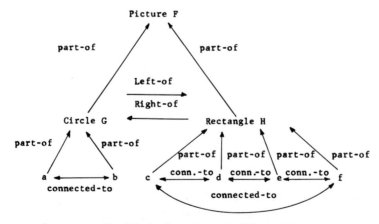

Fig. 1.4. (a) Structural description of the rectangle in Fig. 1.3b. (b) Pattern ⌐ and its structural description.

is shown in Fig. 1.5. Since there is a one-to-one corresponding relation between a linear graph and a matrix, a relational graph can certainly also be expressed as a " relational matrix." In using the relational graph for pattern description, we can broaden the class of allowed relations to include any relation that can be conveniently determined from the pattern. (Notice that (i) the concatenation is the only natural operation for one-dimensional languages, and (ii) a graph, in general, contains closed loops, whereas a tree does not.) With this generalization, we may possibly express richer descriptions than we can with

Fig. 1.5. A relational graph of Picture *F*.

tree structures. However, the use of tree structures does provide a direct channel for adapting the techniques of formal language theory to the problem of compactly representing and analyzing patterns containing a significant structural content. Because of the adaptation of techniques from formal language theory, the syntactic approach is also sometimes called the "linguistic approach." Nevertheless, it is probably more appropriate to consider that the techniques of formal language theory are only tools for the syntactic approach rather than the approach itself. Our motivation for providing efficient structural descriptions and analyses of patterns should not be constrained by the developments in mathematical linguistics.

1.2 Syntactic Pattern Recognition System

A syntactic pattern recognition system can be considered as consisting of three major parts, namely, preprocessing, pattern description or representation, and syntax analysis.† A simple block diagram of the system is shown in Fig. 1.6. The functions of preprocessing include (i) pattern encoding and

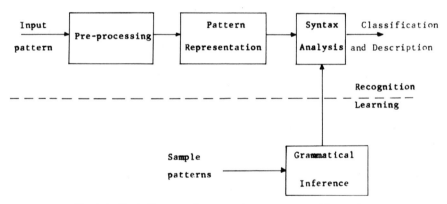

Fig. 1.6. Block diagram of a syntactic pattern recognition system.

approximation; and (ii) filtering, restoration, and enhancement. An input pattern is first coded or approximated by some convenient form for further processing. For example, a black-and-white picture can be coded in terms of a grid (or a matrix) of 0's and 1's, or a waveform can be approximated by its time samples or a truncated Fourier series expansion. In order to make the processing in the later stages of the system more efficient, some sort of "data

---

† The division into three parts is for convenience rather than necessity. Usually, the term "syntactic pattern recognition" refers primarily to the pattern representation (or description) and the syntax analysis.

compression" is often applied at this stage. Then, techniques of filtering, restoration, and/or enhancement are used to clean the noise, to restore the degradation, and/or to improve the quality of the coded (or approximated) patterns. At the output of the preprocessor, presumably, we have patterns of reasonably "good quality." Each preprocessed pattern is then represented by a language-like structure (e.g., a string). This pattern representation process consists of (i) pattern segmentation, and (ii) primitive (feature) extraction. In order to represent a pattern in terms of its subpatterns, we must segmentize the pattern and, in the meantime, identify (or extract) the primitives in it. In other words, each preprocessed pattern is segmentized into subpatterns and pattern primitives based on prespecified syntactic or composition operations; and, in turn, each subpattern is identified with a given set of pattern primitives. At this point, each pattern is represented by a set of primitives with specified syntactic operations.† For example, in terms of the concatenation operation, each pattern is represented by a string of (concatenated) primitives. The decision whether or not the representation (pattern) is syntactically correct (i.e., belongs to the class of patterns described by the given syntax or grammar) will be made by the "syntax analyzer" or "parser." When performing the syntax analysis or parsing, the analyzer can usually produce a complete syntactic description, in terms of a parse or parsing tree, of the pattern, provided the latter is syntactically correct. Otherwise, the pattern is either rejected or analyzed on the basis of other given grammars, which presumably describe other possible classes of patterns under consideration.

Conceptually, the simplest form of recognition is probably "template matching." The string of primitives representing an input pattern is matched against strings of primitives representing each prototype or reference pattern. Based on a selected "matching" or "similarity" criterion, the input pattern is classified in the same class as the prototype pattern that is the "best" match to the input. The hierarchical structural information is essentially ignored. A complete parsing of the string representing an input pattern, on the other hand, explores the complete hierarchical structural description of the pattern. In between, there are a number of intermediate approaches. For example, a series of tests can be designed to test the occurrence or nonoccurrence of certain subpatterns (or primitives) or certain combinations of subpatterns or primitives. The result of the tests (e.g., through a table look-up, a decision tree, or a logical operation) is used for a classification decision. Notice that each test may be a template-matching scheme or a parsing for a subtree representing a subpattern. The selection of an appropriate approach for recognition usually depends on the problem requirement. If a complete

---

† Presumably, more sophisticated systems should also be able to detect the syntactic relations within the pattern.

pattern description is required for recognition, parsing is necessary. Otherwise, a complete parsing could be avoided by using other simpler approaches to improve the efficiency of the recognition process.

In order to have a grammar describing the structural information about the class of patterns under study, a grammatical inference machine is required that can infer a grammar from a given set of training patterns in language-like representations.† This machine's function is analogous to the "learning" process in a decision-theoretic pattern recognition system [1–11, 17–20]. The structural description of the class of patterns under study is learned from the actual sample patterns from that class. The learned description, in the form of a grammar, is then used for pattern description and syntax analysis. (See Fig. 1.6.) A more general form of learning might include the capability of learning the best set of primitives and the corresponding structural description for the class of patterns concerned.

In the following chapters, after a brief introduction to formal languages (Chapter 2), the problems of structural description of patterns (Chapter 3), syntax analysis (Chapter 4), and grammatical inference (Chapter 7) will be discussed more extensively. The problem of preprocessing, which is common to both the decision-theoretic and the syntactic approaches and is often treated as a separate subject from recognition, will be only briefly reviewed in Section 1.3. Various techniques for pattern segmentation will be summarized in Section 1.4.

1.3   Preprocessing Techniques‡

1.3.1   Encoding and Approximation

For the purpose of processing by digital computers, patterns are usually digitized (in time or in space) first, and then represented by more compact forms (data compression) using coding or approximation schemes. Consider that a pattern is a one-dimensional (e.g., waveforms) or two-dimensional (e.g., pictures) function. Based on the following sampling theorems, we know that a given function can be exactly reconstructed by the samples taken at a finite set of points.

Theorem 1.1   Let $f(t)$ be a function whose Fourier transform is zero outside the interval $-W \leq \omega \leq W$. Then $f(t)$ can be exactly reconstructed from samples of its values taken $\frac{1}{2}W$ apart or closer.

---

† At present, this part is performed primarily by the designer.
‡ For more extensive discussions on preprocessing, see the literature [9, 21–28].

Theorem 1.2 Let $f(x, y)$ be a function whose Fourier transform is zero outside a bounded subset B of the plane. Let (a, b) and (c, d) be any points such that, for any two pairs of integers (m, n) and (p, q), the set obtained by shifting B by the amounts $(m(a + c), n(b + d))$ and $(p(a + c), q(b + d))$, respectively, have no point in common. Let (s, t) and (u, v) be such that $as + bt = cu + dv = 2\pi$; $au + bv = cs + dt = 0$. Then $f(x, y)$ can be exactly reconstructed from samples of its values taken at the point $(hs + ku, ht + kv)$, where h and k are integers.

In practice, the Fourier transform of f will usually be close to zero only outside the interval $(-W, W)$ or the subset B. Hence, the reconstruction of f from its samples will not be exact; that is, it is only an approximation of f. Other types of sampling include the use of simple functions of various forms (e.g., constant functions, linear functions, or high-order polynomials) to approximate the pattern function by interpolation between the sampling points.

Quantization is another form of approximation.† In this case, the actual value of the function is approximated by only a finite set of discrete values, called "quantization levels." As in the case of sampling, it is usually simplest to choose quantization levels that are equally spaced. But unequally spaced levels may sometimes be preferable. For example, suppose that, in a picture, the gray levels in a certain range occur more frequently than those in other ranges. In such a case, it might be preferable to use quantization levels that are finely spaced inside this range and coarsely spaced outside it. This method will increase the average accuracy of the quantization without increasing the number of levels, and is known as "tapered quantization." The choice of sampling points or quantization levels need not be prespecified; it can be made to depend on the nature of the patterns being approximated. Of course, the trade-off in this case is that a more sophisticated sampling or quantization scheme would be required.

After a pattern is sampled and quantized, the quantized value at each sample point is usually coded and is used to represent (approximate) the pattern function at the point. Let there be m possible quantization levels. It will take at least $\log_2 m$ bits to represent any given level. If the probability of occurrence of each possible level is known, a more efficient code can be constructed to reduce the average number of bits representing each level. The basic idea in constructing such codes is to use short codes for the commonly occurring quantization levels. We consider this step as falling into the area of coding and information theory and, hence, beyond the scope of this book.

---

† Another possible form of approximation is to approximate a pattern function using truncated series expansion (e.g., Taylor series, orthonormal expansions).

1.3.2   Filtering, Restoration, and Enhancement

Time-invariant and position-invariant (or shift-invariant) operations can be used to "filter" a pattern in order to detect a given pattern in it (template matching), to "restore" a pattern that has been degraded by approximation or other such operations, and to "smooth" or "enhance" a pattern to improve its quality. Such operations can be implemented not only by conventional digital computers, but also by special-purpose electrooptical devices. Let $\mathscr{F}$ be a set of pattern functions and T be a translation (shifting) operation in time or in space defined on $\mathscr{F}$. An operation φ is called time invariant (or position invariant) if it commutes with every T in time (or in space). That is,

$$\varphi(T(f)) = T(\varphi(f)) \tag{1.1}$$

for all $f \in \mathscr{F}$. Operations of this type have the property that their effect on a point of the pattern function does not depend on its time or position in the pattern. Shifting operations, point operations ($\varphi(f)$ at a point depends only on the value of f at the point), and local operations† ($\varphi(f)$ at a point depends only on the values of f in some interval or neighborhood about that point) are operations of this type. Typical examples of time- or position-invariant linear operations are, respectively, the one-dimensional and the two-dimensional Fourier transforms, the Hadamard transforms, and the spread and transfer functions. For pictorial patterns, the second Hadamard transform of a Hadamard-transformed picture will result in the original picture, and the second Fourier transform of a Fourier-transformed picture will result in a rotation of the original picture.

In terms of a two-dimensional continuous case, let

$$\varphi(f(x, y)) = \int_{\infty}^{\infty} \int_{\infty}^{\infty} g_\varphi(x - u, y - v) f(u, v) \, du \, dv = f * g_\varphi \tag{1.2}$$

be the convolution of f and g_φ. The function g_φ is called the point-spread function of φ. The interpretation is that φ, when applied to a one-point pattern function, results in spreading (or blurring or smearing) the one-point nonzero value over a region. Notice that the point-spread function in the one-dimensional case corresponds to the impulse response of φ. Instead of considering a one-point pattern function, the idea of using a one-line pattern function has been suggested. The function has the value 1 along a straight line and zero elsewhere. The result of applying φ to such a function is called a line-spread function of φ. Another possibility is that of a function which is zero

---

† Notice that a point operation is a special case of a local operation, in which the interval or the neighborhood concerned consists of just the point itself.

on one side of a straight line and 1 on the other side; in this case, the result of applying φ is called an edge-spread function of φ. Also, the interpretation of the impulse response of φ can be extended to the case of sinusoidal response; for example, $f(u,\ v) = a + b \cos 2\pi\omega u$ in (1.2). In this case, the result of applying φ is called a transfer function of φ.

It is often necessary to determine how well two patterns match (or are similar to) each other, or to detect if a part of one pattern (a subpattern) matches another pattern. The cross-correlation between two pattern functions f and g provides a simple method of determining whether or not they are identical except for translation and multiplication by a constant. This method can be very useful for pattern or primitive recognition problems in which prototype patterns or templates can be easily determined. The recognition task can be accomplished by cross-correlating the input pattern with each of the templates. Similarly, the problem of finding the pattern f in the pattern g can be solved by computing the quotient of two cross-correlations. Suppose that the pattern g contains a "noisy" version of the pattern f, and we want to find f in g by cross-correlating a template f^* with g. If the noise is additive and independent of f, it is easy to show that the best f^* to use for detecting f is f itself, which is precisely a "matched filter."

Another way to match a pattern with a template is, instead of convolving the pattern with the template, to multiply their corresponding Fourier transforms and to take the inverse Fourier transform of the product. In this case, the matching operation is carried out in the frequency or spatial-frequency domain rather than in the time or space domain. An important application of this filtering technique is to the problem of pattern restoration. Suppose that a pattern has been obtained through some transmission or approximation process that has degraded it. If the degraded pattern can be represented by $f * g$, that is, by a convolution of some g with the original pattern f, then we can "restore" the pattern by taking the inverse Fourier transform of FG/G where FG is the Fourier transform of $f * g$.

In order to suppress noise that may be present in a pattern, the smoothing operation is often used. We can smooth a pattern by simply replacing its value at each point by the average of the values over a neighborhood of the point. This operation can be easily implemented through the convolution of the pattern with a function that has the value $1/B$ inside the neighborhood and 0 outside, where B is the length (in the one-dimensional case) or the area (in the two-dimensional case) of the neighborhood. The size of the neighborhood may be fixed or may vary from point to point. The averaging operation can also be performed on a set of independent copies of a noisy pattern. In addition to the linear operation as the convolution, nonlinear operations can be used for averaging. The simplest class of such operations is that in which averaging is combined with thresholding. Another way of smoothing a pattern is to perform the operation in the frequency or spatial-frequency domain by high-pass or band-pass filtering.

In contrast to smoothing or blurring a pattern, we can "sharpen" or enhance it by clipping, filtering, or differentiation. Since integration (or averaging) can smooth a pattern, a natural approach to sharpening is to perform some sort of differentiation. The use of the gradient and Laplacian has been suggested for the differentiation of pictorial patterns. If we want to enhance a picture in every direction, we can take the derivative in the gradient direction at each point (the direction along which the gray level changes fastest). This operation can be implemented, at least approximately, by convolving the given picture with appropriate templates or by using a gradient-matched filter. Edges or lines in a particular direction can be emphasized, while all others are alternated by performing differentiation in the orthogonal direction, which can also be implemented by convolution. The Laplacian is a useful combination of derivatives, which is proportional to the difference between the gray level at the point and the average gray level in an annulus centered at the point. The Laplacian operation can be approximated by convolving the picture with a template having a positive peak that is surrounded by a negative annular valley, with the value chosen so that the integral over the template is zero.

1.4 Pattern Segmentation

1.4.1 Picture Segmentation and Primitive Extraction†

There is no universal method of segmenting a picture into subpictures. Different types of subpictures can be useful, depending on the type of description that is required. A basic method for singling out a subpicture from a picture is to threshold the given picture. We can define subpictures by simply thresholding the gray level of the picture itself. The subpicture can be defined in terms of the set of points in the picture at which the gray levels are greater or less than a threshold, or are between two thresholds. For example, in the recognition of characters or line drawings, the characters or drawings are "black" and the paper or background "white"; a simple thresholding of the picture itself would separate the black pattern from the white background. The thresholding operation can also be combined with other preprocessing operations, such as smoothing, sharpening, and matched filtering.

An important problem in using the thresholding operation is the selection of the threshold value. It is sometimes possible to select a good threshold by

† For more extensive discussions on picture segmentation, see the literature [21, 29, 30].

examining the frequency distribution of gray levels of the picture. For example, if the approximate area of the desired subpicture is known, we can choose the lowest (or the highest) threshold such that the area of the subpicture at which it is exceeded is below (or above) a prespecified proportion of the area of the picture (the p-tile method). On the other hand, if the subpicture contains a range of gray levels different from those which occur in the rest of the picture, the distribution will typically show a peak corresponding to the levels in the subpicture; and we can extract the subpicture by selecting thresholds at the bottoms of the valleys on the two sides of this peak (the mode method). Another method for selecting the threshold is to test points for membership in the subpicture sequentially. This procedure makes it possible to vary the threshold value in a manner dependent on the nature of the points that have already been accepted. The result is, of course, dependent on the sequence in which the points are tested.

Another approach to extracting the subpictures is "boundary (or contour) following and tracking." In character or line patterns, the boundary-following method would make it easy to obtain information about the pattern's shape while extracting. A simple way of following boundaries in a binary digital picture is to search the picture systematically until a pair of adjacent picture elements or cells having different values is found. Suppose that we move along the boundary between the elements keeping the 1 (black) on the right, say, until we reach a "crossroads" with the following situation.

Depending upon the values of x and y, we choose the "next" boundary point according to the following rule.

x	y	Next boundary point
0 or 1	0	Turn right
0	1	Go straight
1	1	Turn left

This procedure will follow clockwise around the boundary of the connected element of 1's. The boundary-following algorithm can be generalized to a gray-scale digital picture if a criterion for the existence of a boundary is given. (For example, a boundary between two elements exists when the elements' gray levels differ by more than some threshold value.) Boundaries can be "tracked," even if they have gaps (e.g., in fingerprint patterns and in bubble chamber and spark chamber pictures), by continuing in the previous direction whenever the boundary comes to an end and going on for a distance corresponding to the largest allowable gap.

The boundary-following procedure can be modified such that it will "follow" the border elements rather than the "cracks" between these elements and their neighboring 0's. Let Sp be a subpicture of a (digital) picture which is connected (has only one component), and let $\overline{Sp}$ denote the complement of Sp, that is, the set of picture elements not in Sp. Let B be the set of border elements of Sp, and let $x_0 = (i_0, j_0)$ be the initial element of B. Since (i_0, j_0) is a border element of Sp, at least one of the four (horizontal and vertical) neighboring elements, that is, $(i_0 + 1, j_0)$, $(i_0 - 1, j_0)$, $(i_0, j_0 + 1)$, $(i_0, j_0 - 1)$, is in $\overline{Sp}$. Denote this element by $y_0{}^1$. (Choose any one if more than one is in $\overline{Sp}$.) Label the eight neighbors (horizontal, vertical, and diagonal neighbors) of (i_0, j_0) in counterclockwise order (keeping Sp on the left), starting with $y_0{}^1$, as $y_0{}^1, y_0{}^2, \ldots, y_0{}^8$. If none of $y_0{}^3, y_0{}^5, y_0{}^7$ is in Sp, then x_0 is the sole element of Sp. Otherwise, let y_0^{2k+1} be the first of these y_0's that is in Sp. The next border element x_1 of B is determined according to the rule:

$$x_1 = \begin{cases} y_0^{2k+1} & \text{if } y^{2k} \text{ is in } \overline{Sp} \\ y_0^{2k} & \text{if } y_0^{2k} \text{ is in } Sp \end{cases} \tag{1.3}$$

$i-1, j+1$	$i, j+1$	$i+1, j+1$
$i-1, j$	i, j	$i+1, j$
$i-1, j-1$	$i, j-1$	$i+1, j-1$

To find the next border element, take the neighbor of x_1 that is in $\overline{Sp}$ (y_0^{2k} or y_0^{2k-1}) as $y_1{}^1$ and repeat the same procedure.

Once a subpicture has been extracted from a picture, new operations can be performed on the subpicture to derive new subpictures with special properties. For example, if we have used the thresholding operation to extract a subpicture from a picture (which may contain solid regions, lines, or contours), we can then apply a boundary- or border-following operation to derive picture boundaries (new subpictures) with the property of "connectivity." By proper labeling we can single out (give a "name" or a "code" to) each component of a given subpicture, say, by assigning to it a positive integer in such a way that all elements in a given connected component of the subpicture receive the same label, while elements in different components receive different labels. Another operation involving connectivity is the "shrinking" and "thinning" of a subpicture. If a subpicture has many small connected components, we may want to simplify the picture by shrinking each component down to a single element. Similarly, if a subpicture is elongated everywhere, we may want to thin it into a line drawing. To replace components by single elements, we can simply label them and then suppress all but one instance of each label. To thin a subpicture, we can apply a shrinking operation in order to remove successive layers of border elements from it. (Notice that additional conditions may be required to prevent the subpicture from becoming disconnected.)

Distance functions can be used to derive various types of subpictures from a given subpicture. Let d be a distance function defined for a pair of elements which is positive definite, symmetric, and satisfies the triangle inequality. With different distance functions selected, the set of elements (l, m) at distances from (i, j) less than or equal to a threshold value r differs in shape, as shown in Fig. 1.7. The set of elements from which the distance to $\overline{Sp}$ is a local

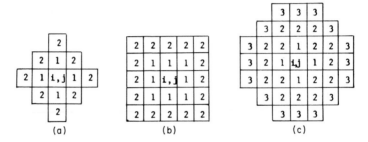

Fig. 1.7. Distances from element (i, j). (a) Diamond, $r = 2$; (b) square, $r = 2$; (c) octagon, $r = 3$.

maximum constitutes a sort of "skeleton" of Sp. Skeletons provide an alternative approach to thinning; they can also be used to extract information about the shape of a subpicture. Other operations involving distance functions include "isolation," the "expansion" (fattening) and "elongation" of a subpicture, and the calculation of the sum of gray levels of the elements at a distance r from element (i, j), among others.

We can also derive subpictures in terms of the direction and shape of a given subpicture. For a line pattern, we can define new subpictures by labeling each point of the curve with the approximate direction of the curve at the point. "Break points" of a curve or the boundary or border of a subpicture can be extracted for segmentation in terms of position extrema (the highest, lowest, rightmost, or leftmost points of the curve), points of reflection on the curve, or points at which their curvatures are local maxima or change abruptly. We can also take as a new subpicture the set of points that are hidden by a given subpicture when we look toward it from far away in a given direction (a "shadow"). Moreover, the set of points at which a ray in a given direction intersects the given subpicture a given number of times can also be used as a new subpicture. For shape features, we can extract "strokes," "bays," and the like in a line pattern. These subpictures involving shape can be detected by matching the given subpicture with two-dimensional templates or with mathematical functions describing the shape features. Straight lines can also be detected by applying a "point-to-line" or "point-to-curve" transformation. Similarly, the presence of other simple shapes, such as circular arcs, can often be detected by performing appropriate coordinate transformations.

After the subpictures or primitives are explicitly singled out from a picture, we can calculate various picture properties to further characterize each subpicture or primitive. These properties can be expressed numerically, as functions, or as logical statements (predicates). The role that these properties play may be regarded as primarily "semantic" rather than syntactic. For pictorial patterns, geometric or topological properties are often very useful. The number of connected components in a picture (i.e., "connectivity") can be easily counted through a component labeling method. Let S be a connected subpicture. The number of components in $\bar{S}$ is called the "order of connectivity" of S, which is equal to the number of "holes" in S plus one. (S is simply connected if and only if its order of connectivity is 1.) If we know that every connected component of S is simply connected, we can calculate the connectivity of S by computing its Euler number or "genus." The Euler number of S, E_S, is defined as the number of components C of S minus the number of holes H in S. The Euler number of a digital picture S can be calculated from

$$E_S = C - H = N_S - Z_S + T_S \tag{1.4}$$

where N_S is the number of elements in S, Z_S the number of horizontally (or vertically) adjacent pairs of elements in S, and T_S the number of two-by-two squares of horizontally and vertically adjacent elements in S. Equation (1.4) is a "dual" of a more familiar expression for polygonal networks:

$$E = N_v - N_e + N_f \tag{1.5}$$

where N_v, N_e, N_f are the number of vertices, edges, and faces, respectively. Other geometric properties often used for describing pictures include length and curvature of a curve; area, perimeter, height, width, and extent of a subpicture;† the number of position extrema, points of reflection, concavities

---

† The area of a digital subpicture is just the total number of elements in the subpicture; the perimeter can be taken as being approximately proportional to the area of the subpicture's border. The height or width of a subpicture can be measured as the distance between the highest and the lowest horizontal lines, or the rightmost and the leftmost vertical lines. Similarly, the "extent" of a subpicture is called its "diameter."

("loops"), curvature maxima ("angles") in a subpicture; etc. It is known that a subpicture can be completely determined by specifying its boundary. One way to specify the boundary is to describe it by an equation or a set of equations. The boundary of a subpicture can also be described by its "intrinsic equation," which represents its curvature or slope as a function of arc length. Other functional properties include frequency distributions of slopes or curvatures of the subpicture's boundary and of distances from the subpicture to itself.

1.4.2   Segmentation of Continuous Speech

Connected speech is produced by a nearly continuous motion of the vocal apparatus from sound to sound. Written equivalents of such speech sounds are usually represented by a string of linguistic elements called phonemes. This process raises the problem of segmentation of the acoustical continuum of speech sounds into discrete segments that can then be associated with specific phonemes. Humans are able to perform the segmentation function in a natural way, although, depending on their culture, they may segment a given speech sound differently. Machine segmentation of a continuous speech wave can be performed in the time, frequency, or time–frequency (spectrograph) domain [31–36]. Existing machine segmentation techniques are primarily derived in an ad hoc way. A time-domain speech segmentation technique developed by Reddy [33, 34] is briefly described in this section.

The speech wave is first divided into a succession of "minimal segments." The duration of each segment is chosen small enough so that significant changes do not occur within the segment. In a typical example, the duration of each minimal segment is 10 msec, and each contains 100 points at 100-μsec intervals. Let the amplitudes of the speech waveform at the n equidistant points ($n = 100$) within a minimal segment be represented by an n-dimensional vector $X = [x_1, x_2, \ldots, x_n]$. The intensity level I of a minimal segment is defined as

$$I = \max_{1 \leq i \leq n} |x_i| \tag{1.6}$$

Let Y be an $(n - 1)$-dimensional vector such that

$$y_i = \begin{cases} 1 & \text{if } x_i x_{i+1} < 0 \\ 0 & \text{otherwise} \end{cases} \tag{1.7}$$

The number of "zero crossings" of a minimal segment Z is then given by

$$Z = \sum_{i=1}^{n-1} y_i \tag{1.8}$$

Successive minimal segments with similar characteristics are grouped together to form larger segments. Intensity levels and zero crossings are calculated for each minimal segment. For all the vowels, voiced fricatives, liquids, nasals, and the silence preceding stops, the intensity level has been found to be about the same during major portions of the utterance of the phoneme except at boundaries. The variability was found to be greater in terms of zero crossings. Thus, intensity levels can be used to group together acoustically similar minimal segments and zero crossings to resolve the ambiguities. Two adjacent segments with intensity levels I_1 and I_2 are considered to be acoustically similar if there exists a tolerance interval T such that $I_1 \in T$ and $I_2 \in T$ where T is defined to be the closed interval $[I - t, I + t]$, and $t = \max \{1, R_I/8\}$. R_I is the range of I, and for six-bit samples, $R_I = [0, 32]$. A "sustained segment" is a segment of no less than 30-msec duration and is a union of acoustically similar segments. A "transitional segment" is defined as any segment that is not a sustained segment. By inspecting a group of successive minimal segments of no more than 200-msec duration, a distribution of acoustically similar segments within the tolerance intervals of each of the possible intensity levels is determined. The most likely interval at which a sustained segment may be found is taken to be the one that contains the largest number of minimal segments. Using the criterion for acoustical similarity, an attempt is made to group as many segments as possible within the tolerance interval to form a sustained segment. If the attempt fails, then the next most likely interval is used, and so on until success is achieved or it is determined that no sustained segment can be found. A typical result of an experiment conducted by Reddy is shown in Fig. 1.8.

After the segmentation of a continuous speech wave is completed, each segment is associated with a phoneme or a transitional segment between two phonemes. Those segments that are not rejected as being transitions are then classified into phonemes. Since phonemes are considered to be the basic linguistic elements (primitives) of speech sounds, phoneme classification procedures are, in general, nonsyntactic (decision-theoretic or heuristic, etc.). The features extracted for phoneme classification are generally the time-domain and/or frequency-domain parameters (e.g., duration, intensity, zero crossings, pitch, formant frequencies and amplitudes) and their transformations. If syntactic rules (grammar) can be obtained to describe the speech patterns under study, then the syntactic approach can be applied (Appendix F).

1.5 Remarks on Syntactic Approach versus Decision-Theoretic Approach

Some successful results have been obtained in applying syntactic approaches to character recognition, chromosome analysis, bubble chamber picture identification, and recognition of two-dimensional mathematical expressions

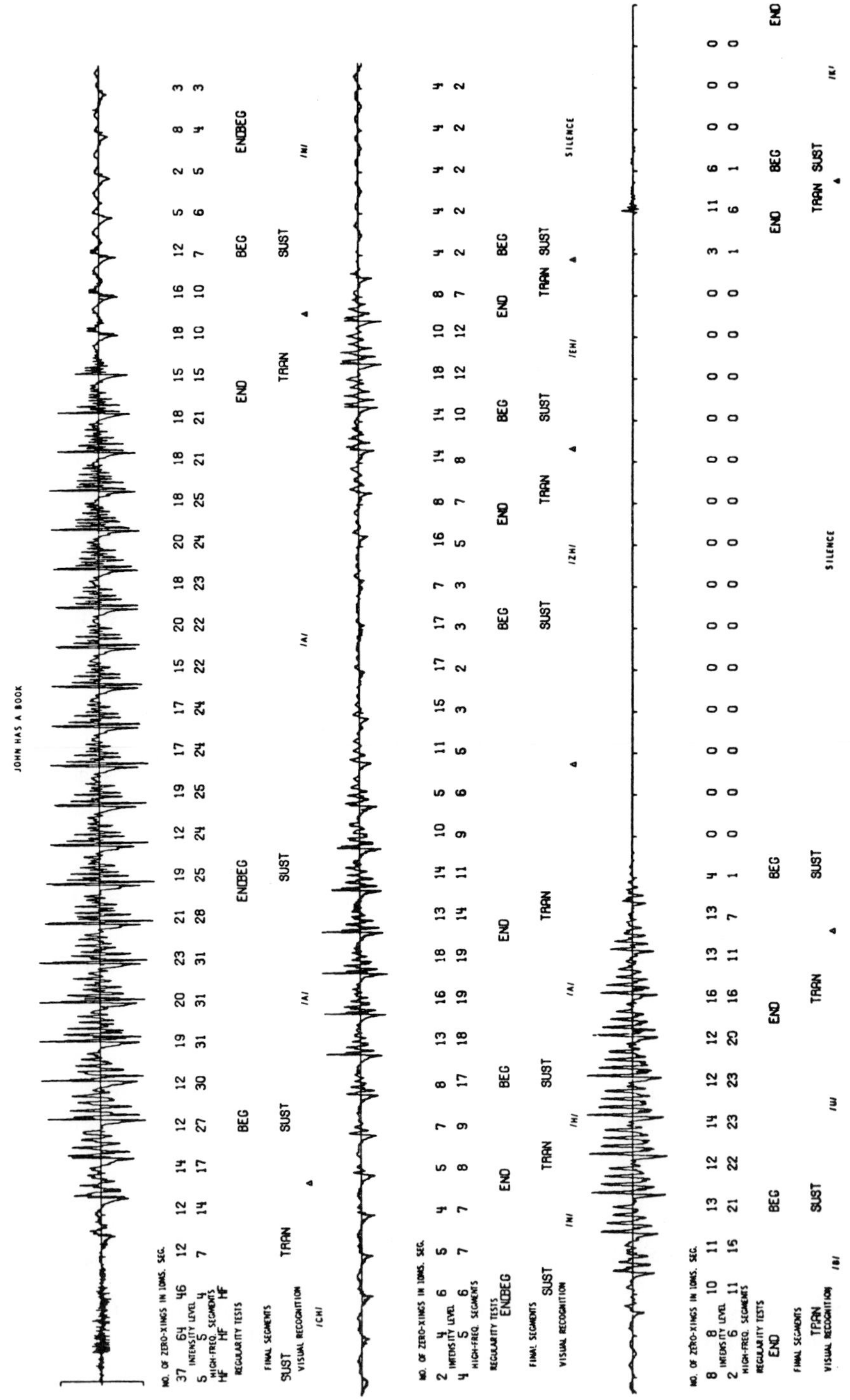

Fig. 1.8. Segmentation of the sound "John has a book." (From Reddy [34].)

(e.g., see the literature [37–41]). However, the dichotomy of syntactic and decision-theoretic approaches appears to be convenient only from the viewpoint of theoretical study. In other words, the division of the two approaches is sometimes not clear-cut, particularly in terms of practical applications. Many preprocessing techniques are useful in both approaches, apparently because the selection of a particular preprocessing technique now depends more on the type of the patterns (pictorial or waveform, highly noisy, degraded or not, etc.) under study than the recognition approach being applied. The feature extraction and selection problem in the decision-theoretic approach and the primitive extraction and selection problem in the syntactic approach are similar in nature [29, 30] except that the primitives in the syntactic approach represent subpatterns and, on the other hand, the features in the decision-theoretic approach may be any set of numerical measurements taken from the pattern. As a matter of fact, it appears that there is no distinction between the two approaches at this level,† since it is intended that the pattern primitives selected should contain no significant syntactic information with respect to the recognition problem. For example, in the recognition of two-dimensional mathematical expressions, the approach used is primarily syntactic and the pattern primitives are the characters and mathematical symbols [40, 41]. It may well be that a decision-theoretic technique is used to recognize pattern primitives and a syntactic method is used for recognition of mathematical expressions. A similar situation may also arise in the recognition of continuous speech. That is, the relations between a spoken sentence and its phrases and words (subpatterns of a sentence), and between a spoken word and its phonemes, syllables, or sounds may be regarded as primarily syntactic [42–44].‡ However, the recognition of the pattern primitives (in this case, e.g., the phonemes, syllables, or sounds) may be more effectively accomplished by applying decision-theoretic techniques [45–47].

From the foregoing discussion, it appears that, with respect to practical pattern recognition problems, the syntactic and decision-theoretic approaches in many cases complement each other. When explicit structural information about the patterns is not considered important and the problem is primarily one of classification rather than of classification and description, then there appears to be no real need for using syntactic methods. When, on the other hand, patterns are very rich in structural information and the recognition problem requires classification and description, then the syntactic approach seems to be necessary. In many pattern recognition applications, the problem

---

† Refer to Section 1.4 for some of the primitive extraction techniques.

‡ This is, of course, under the assumption that the segmentation problem can be successfully solved.

essentially falls between these two extreme cases. In other words, some sort of structural information about patterns is important and should be utilized, but simple and abstract pattern primitives may not be easy to extract, particularly for noisy and distorted patterns. Since, in general, pattern primitives are defined more or less in terms of local properties of a pattern (which are more sensitive to noise and distortion) and contain little syntactic information, they probably can be more effectively recognized by decision-theoretic techniques. After the pattern primitives are extracted, a syntactic method would be applied for the recognition of the whole pattern. An appropriate utilization of the combination of these approaches may result in an efficient, practical system. At the lower level (in terms of the hierarchical structure of the patterns), a decision-theoretic technique is used to recognize pattern primitives. Based on this principle, the pattern primitives selected for a given recognition problem can be defined as the subpatterns that can be recognized by decision-theoretic techniques. Then the higher-level structural information of the patterns will be expressed in terms of these subpatterns. In this case (a two-level scheme), the primitives, as long as they can be recognized, are not necessarily very simple and abstract. Consequently, the resulting syntactic description may turn out to be quite simple, and the recognition problem can be solved efficiently by applying a syntactic method.

When the patterns are noisy and distorted, the appearance of two or more different patterns (patterns belonging to two or more different classes) may appear to be the same.† In terms of the syntactic approach, this represents the situation where one pattern has two or more different structural descriptions, which, in turn, results in the so-called ambiguity in formal language theory. This ambiguity in structural descriptions cannot be discriminated by normal syntax analysis procedures. In such circumstances, the discrimination among different structural descriptions may be carried out by decision-theoretic techniques from the statistical information about the pattern noise and distortion. This leads to the extension of formal languages to include stochastic languages for syntactic pattern recognition [48].‡ Both statistical and structural information about the patterns is conveyed by stochastic grammars, and the classification and description of patterns will be performed by stochastic syntax analysis. The statistical information, in general, would also speed up the syntax analysis. After a review of stochastic grammars and languages in Chapter 5, the application of stochastic languages to syntactic pattern recognition will be discussed in Chapter 6.

† This, of course, depends partially on the effectiveness of preprocessing techniques.
‡ The use of transformational grammar has also recently been proposed as an alternative approach by which to describe deformed patterns [49].

References

1. K. S. Fu, *Sequential Methods in Pattern Recognition and Machine Learning.* Academic Press, New York, 1968.
2. G. S. Sebestyen, *Decision Processes in Pattern Recognition.* Macmillan, New York, 1962.
3. N. J. Nilsson, *Learning Machines-Foundations of Trainable Pattern-Classifying Systems.* McGraw-Hill, New York, 1965.
4. J. M. Mendel and K. S. Fu, eds., *Adaptive, Learning and Pattern Recognition Systems: Theory and Applications.* Academic Press, New York, 1970.
5. W. Meisel, *Computer-Oriented Approaches to Pattern Recognition.* Academic Press, New York, 1972.
6. K. Fukunaga, *Introduction to Statistical Pattern Recognition.* Academic Press, New York, 1972.
7. E. A. Patrick, *Fundamentals of Pattern Recognition.* Prentice-Hall, Englewood Cliffs, New Jersey, 1972.
8. H. C. Andrews, *Introduction to Mathematical Techniques in Pattern Recognition.* Wiley, New York, 1972.
9. R. O. Duda and P. E. Hart, *Pattern Classification and Scene Analysis.* Wiley, New York, 1973.
10. C. H. Chen, *Statistical Pattern Recognition.* Hayden, Washington, D.C., 1973.
11. J. Sklansky. ed., *Pattern Recognition: Introduction and Foundations.* Dowden, Hutchinson and Ross, Inc., Stroudsburg, Pennsylvania, 1973.
12. K. S. Fu and P. H. Swain, On syntactic pattern recognition. In *Software Engineering* (J. T. Tou, ed.), Vol. 2. Academic Press, New York, 1971.
13. W. F. Miller and A. C. Shaw, Linguistic methods in picture processing—A survey. *Proc. AFIPS Fall Joint Comput. Conf., San Francisco, 1968,* pp. 279–290.
14. R. Narasimhan, *A Linguistic Approach to Pattern Recognition.* Rep. 121. Digital Comput. Lab., Univ. of Illinois, Urbana, 1962.
15. *Pattern Recognition* **3** (1971); **4** (1972). Special issues on syntactic pattern recognition.
16. N. V. Zavalishin and I. B. Muchnik, Linguistic (structural) approach to the pattern recognition problem. *Automat. i Telemeh.* pp. 86–118 (1969).
17. Ya. Z. Tsypkin, *Foundations of the Theory of Learning System.* Nauka, Moscow, 1970.
18. M. A. Aiserman, E. M. Braverman, and L. I. Rozonoer, *Potential Function Method in Theory of Learning Machines.* Nauka, Moscow, 1970.
19. K. S. Fu, ed., *Pattern Recognition and Machine Learning.* Plenum, New York, 1971.
20. A. G. Arkadev and E. M. Braverman, *Learning in Pattern Classification Machines.* Nauka, Moscow, 1971.
21. A. Rosenfeld, *Picture Processing by Computer.* Academic Press, New York, 1969.
22. J. K. Hawkins, Image processing: A review and projection. In *Automatic Interpretation and Classification of Images* (A. Grasselli, ed.). Academic Press, New York, 1969.
23. H. C. Andrews, *Computer Techniques in Image Processing.* Academic Press, New York, 1970.
24. *Pattern Recognition* **2**, (1970). Special issue on image enhancement.
25. T. S. Huang, W. F. Schreiber, and O. J. Tretiak, Image processing. *Proc. IEEE* **59**, 1586–1609 (1971).
26. K. Preston, Jr., *Coherent Optical Computers.* McGraw-Hill, New York, 1972.
27. *IEEE Trans. Computers.* **C-21** (1972). Special issue on two-dimensional digital signal processing.

28. *IEEE Proc.* **60** (1972). Special issue on digital picture processing.
29. *IEEE Trans. Computers* **C-20** (1971). Special issue on feature extraction and selection in pattern recognition.
30. *Pattern Recognition* **3** (1971). Special issue on feature extraction.
31. T. Sakai and S. Doshita, The automatic speech recognition system for conversational sound. *IEEE Trans. Electron. Comput.* **EC-12**, 835–846 (1963).
32. P. N. Sholtz and R. Bakis, Spoken digit recognition using vowel-consonant segmentation. *J. Acoust. Soc. Amer.* **34**, 1–5 (1962).
33. D. R. Reddy, Segmentation of speech sounds. *J. Acoust. Soc. Amer.* **40**, 307–312 (1966).
34. D. R. Reddy, *An Approach to Computer Speech Recognition by Direct Analysis of the Speech Wave.* Tech. Rep. No. CS 49. Comput. Sci. Dept., Stanford Univ., Stanford, California, September 1, 1966.
35. A. L. Nelson, M. B. Hersher, T. B. Martin, H. J. Zadell, and J. W. Falter, Acoustic recognition by analogue feature-abstraction techniques. In *Models for the Perception of Speech and Visual Form* (W. Wathen-Dunn, ed.), pp. 428–440. MIT Press, Cambridge, Massachusetts, 1967.
36. C. C. Tappert, A preliminary investigation of adaptive control of the interaction between segmentation and continuous speech. *IEEE Trans. Syst. Man, and Cybernet.* **SMC-2**, 66–72 (1972).
37. H. Genchi, K. I. Mori, S. Watanabe, and S. Katsuragi, Recognition of handwritten numerical chapters for automatic letter sorting. *Proc. IEEE* **56** 1292–1301 (1968).
38. R. S. Ledley, L. S. Rotolo, T. J. Golab, J. D. Jacobsen, M. D. Ginsburg, and J. B. Wilson, FIDAC: Film input to digital automatic computer and associated syntax-directed pattern-recognition programming system. In *Optical and Electro-Optical Information Processing* (J. T. Tippet, D. Beckowitz, L. Clapp, C. Koester, and A. Vanderburgh, Jr., eds.). MIT Press, Cambridge, Massachusetts, 1965.
39. A. C. Shaw, *The Formal Description and Parsing of Pictures.* Ph.D. Thesis, Rep. SLAC-84, UC-32, Stanford Univ., Stanford, California, April 1968.
40. R. H. Anderson, *Syntax-Directed Recognition of Hand-Printed Two-Dimensional Mathematics.* Ph.D. Thesis, Harvard Univ., Cambridge, Massachusetts, January 1968.
41. S. K. Chang, A method for the structural analysis of two-dimensional mathematical expressions. *Information Sci.* **2** 253–272 (1970).
42. W. A. Lea, An approach to syntactic recognition without phonemics. *Proc. Conf. Speech Commun. Process. Newton, Massachusetts, April 24–26, 1972,* Conf. Record.
43. K. P. Li, G. W. Hughes, and T. B. Snow, Segment classification in continuous speech. *IEEE Trans. Audio Electroacoust.* **AU-21**, 50–57 (1973).
44. R. De Mori, A descriptive technique for automatic speech recognition. *IEEE Trans. Audio Electroacoust.* **AU-21**, 89–100 (1973).
45. L. C. W. Pols, Real-time recognition of spoken words. *IEEE Trans. Computers* **C-20**, 972–978, (1971).
46. D. G. Bobrow and D. H. Klatt, A limited speech recognition system. *Proc. AFIPS Fall Joint Comput. Conf., San Francisco 1968,* **33**, pp. 305–318.
47. T. G. von Keller, An on-line recognition system for spoken digits. *J. Acoust. Soc. Amer.* **49**, Pt. 2, 1288–1296 (1971).
48. K. S. Fu, Syntactic pattern recognition and stochastic languages. In *Frontiers of Pattern Recognition* (S. Watanabe, ed.). Academic Press, New York, 1972.
49. L. Kanal and B. Chandrasekaran, On linguistic, statistical and mixed models for pattern recognition. In *Frontiers of Pattern Recognition* (S. Watanabe, ed.). Academic Press, New York, 1972.

50. D. O. Clayden, M. B. Clowes, and J. R. Parks, Letter recognition and segmentation of running text. *Information and Control* 9, 246–264 (1966).
51. T. Sakai, M. Nagao, and S. Fuhibayashi, Line extraction and pattern detection in a photograph. *Pattern Recognition* 1, 233–248 (1969).
52. A. Herskovits and T. O. Binford, *On Boundary Detection*. AI Memo. No. 183, Project MAC. MIT, Cambridge, Massachusetts, July 1970.
53. R. L. Hoffman and J. W. McCullough, Segmentation methods for recognition of machine printed characters. *IBM J. Res. Develop.* 15, 153–165 (1971).
54. M. D. Kelly, Edge detection in pictures by computer using planning. In *Machine Intelligence 6* (B. Meltzer and D. Michie, eds.), pp. 397–409. Edinburgh Univ. Press, Edinburgh, 1971.
55. A. Rosenfeld and M. Thurston, Edge and curve detection for visual scene analysis. *IEEE Trans. Computers* C-20, 562–569 (1971).
56. R. Stefanelli and A. Rosenfeld, Some parallel thinning algorithms for digital pictures. *J. Assoc. Comput. Mach.* 18, 255–264 (1971).
57. J. L. Flanagan, *Speech Analysis, Synthesis and Perception*. Academic Press, New York, 1965.
58. M. A. Aiserman, Remarks on two problems concerned with pattern recognition. In *Methodologies of Pattern Recognition* (S. Watanabe, ed.). Academic Press, New York, 1969.
59. L. Uhr, ed., *Pattern Recognition: Theory, Experiment, Computer Simulations, and Dynamic Models of Form Perception and Discovery*. Wiley, New York, 1966.
60. L. N. Kanal, ed., *Pattern Recognition*. Thompson Book Co., Washington, D.C., 1968.
61. P. A. Kolers and M. Eden, eds., *Recognizing Patterns: Studies in Living and Automatic Systems*. MIT Press, Cambridge, Massachusetts, 1968.
62. V. A. Kovalevsky, *Character Readers and Pattern Recognition*. Spartan Books, Washington, D.C., 1968.
63. K. Noda *et al.*, *ASPET/70*. Tech. Rep. Electrotech. Lab., Ministry of Int. Trade, Tokyo, Japan, 1970.
64. *Pattern Recognition* 2 (1970). Special issue on character recognition.
65. R. S. Ledley, Practical problems in the use of computers in medical diagnosis. *Proc. IEEE* 57, 1900–1918 (1969).
66. K. S. Fu, D. A. Landgrebe, and T. L. Phillips, Information processing of remotely sensed agricultural data. *Proc. IEEE* 57 639–653 (1969).
67. I. T. Turbovich, V. G. Gitis, and V. K. Maslov, *Pattern Identification*. Nauka, Moscow, 1971.
68. N. G. Zagoruyko, *Recognition Methods and Their Application*. Radio Sovetskoe, Moscow, 1972.
69. G. C. Cheng, R. S. Ledley, D. K. Pollock, and A. Rosenfeld, *Pictorial Pattern Recognition*. Thompson Book Co., Washington, D.C., 1968.
70. A. Rosenfeld, Picture processing by computer. *Comput. Surveys* 1, 3, 146–176 (1969).
71. E. M. Braverman, ed., *Automatic Analysis of Complex Images*. Mir, Moscow, 1969.
72. A. Rosenfeld, Picture automata and grammars: An annotated bibliography. *Proc. Symp. Comput. Image Process. Recognition, Univ. of Missouri, Columbia, August 24–26, 1972*.
73. M. M. Bongard, *Pattern Recognition* (Engl. transl.). Spartan Books, Washington, D.C. 1970 (Nauka, Moscow, 1967).
74. *IEEE Proc.* (1972). Special issue on digital pattern recognition—L. Harmon, guest ed.
75. L. L. Myasnikov and E. N. Myasnikova, *Automatic Recognition of Acoustic Patterns*. Energinya, Moscow, 1970.

76. G. I. Tzemel, *Identification of Speech Signals.* Nauka, Moscow, 1971.
77. N. G. Zagoruyko and T. I. Zasgovskaya, *Pattern Recognition in Social Studies.* Nauka, Novosibirsk, 1968.
78. U. Grenander, Foundation of pattern analysis. *Quart. Appl. Math.* **27**, 1–55 (1969).
79. U. Grenander, A unified approach to pattern analysis. *Adv. Comput.* **10** (1970).
80. H. G. Barrow and R. J. Popplestone, Relational descriptions in picture processing. In *Machine Intelligence 6,* (B. Meltzer and D. Michie, eds.), pp. 377–396. Edinburgh Univ. Press, Edinburgh, 1971.
81. D. A. O'Handley, The reality of robots. *Symp. Automat. Contr., Milwaukee, Wisconsin, March 10, 1973.*
82. B. S. Lipkin and A. Rosenfeld, eds., *Picture Processing and Psychopictorics.* Academic Press, New York, 1970.
83. M. A. Fischler and R. A. Elschlager, The representation and matching of pictorial structures. *IEEE Trans. Computers* **C-22**, 67–92 (1973).
84. L. D. Menninga, A syntax-directed approach to pattern recognition and description. *Proc. AFIPS Fall Joint Comput. Conf. 1971,* **38**, pp. 145–151.
85. T. Pavlidis and G. S. Fang, A segmentation technique for waveform classification. *IEEE Trans. Computers* **C-21**, 901–904 (1972).
86. T. Sakai, M. Nagao, and H. Matsushima, Extraction of invariant picture substructures by computer. *Comput. Graphics and Image Process.* **1**, 81–96 (1972).
87. E. Wong and J. A. Steppe, Invariant recognition of geometric shapes. In *Methodologies of Pattern Recognition* (S. Watanabe, ed.), Academic Press, New York, 1969.
88. W. Wathen-Dunn, ed., *Models for the Perception of Speech and Visual Forms.* MIT Press, Cambridge, Massachusetts, 1967.
89. C. T. Zahn, A formal description for two-dimensional patterns. *Proc. Int. Joint Conf. Artificial Intelligence, 1st, Washington, D.C., May 1969.*
90. K. M. Sayre, *Recognition: A Study in the Philosophy of Artificial Intelligence.* Univ. of Notre Dame Press, Notre Dame, Indiana, 1965.
91. P. W. Becker, *Recognition of Patterns.* Polyteknisk Forlag, Copenhagen, 1968.
92. J. R. Ullman, *Pattern Recognition Techniques.* Butterworth, London, 1973.
93. R. B. Banerji, A Language for pattern recognition. *Pattern Recognition* **1**, 63–74 (1968).
94. L. Uhr, *Pattern Recognition, Learning and Thought: Computer Programmed Models of High Mental Processes.* Prentice-Hall, Englewood Cliffs, New Jersey, 1973.

Chapter 2

Introduction to Formal Languages

2.1 Introduction

The initial investigation [1] of the mathematical structure of languages was aimed at trying to understand the basic properties of natural languages. It was found that phrase-structure grammar with a set of rewriting rules, which is essentially quite powerful, can be used as a method for describing languages. This concept was then formalized [2] and developed by Chomsky [3, 4], Bar-Hillel and their associates [5, 6]. In late 1960, it was discovered that the "ALGOL-like" languages defined by Backus normal form are identical with the context-free languages that are generated by a spiecial class of phrase-structure grammars. This finding opens the possibility of investigating programming languages from theoretical points of view instead of by heuristic approaches alone. Since then, extensive research has been done by those concerned with either natural languages or programming languages. Different kinds of grammars have been proposed and their properties investigated. A class of recognition devices with automata structure has also been suggested for use in considering the interaction between recognition devices and the languages generated by a certain grammar. Recently, Aho and Ullman [7] and Hopcroft and Ullman [8] provided unified expositions in which the significant results concerning the theory of phrase-structure languages and their relation to automata were described. In this chapter, basic concepts and properties of phrase-structure languages are briefly reviewed. The relation between phrase-structure languages and automata is also discussed.

2.2 Languages and Phrase-Structure Grammars

The concepts of phrase-structure grammars originated in the parsing of a simple English sentence. Before defining language in a formal sense, consider the parsing of the sentence, "The girl walks gracefully." Here, "the girl" is a noun phrase used as the subject, and "walks gracefully" is a verb phrase which consists of the verb "walks" and the adverb "gracefully." Conversely, this sentence is formed by performing the following series of steps:

Step 1 ⟨sentence⟩.
Step 2 ⟨noun phrase⟩ ⟨verb phrase⟩.
Step 3 ⟨adjective⟩ ⟨noun⟩ ⟨verb phrase⟩.
Step 4 The ⟨noun⟩ ⟨verb phrase⟩.
Step 5 The girl ⟨verb phrase⟩.
Step 6 The girl ⟨verb⟩ ⟨adverb⟩.
Step 7 The girl walks ⟨adverb⟩.
Step 8 The girl walks gracefully.

This set of steps, in turn, can be described by the following rewrite rules:

⟨sentence⟩ → ⟨noun phrase⟩ ⟨verb phrase⟩
⟨noun phrase⟩ → ⟨adjective⟩ ⟨noun⟩
⟨verb phrase⟩ → ⟨verb⟩ ⟨adverb⟩
⟨adjective⟩ → the
⟨noun⟩ → girl
⟨verb⟩ → walks
⟨adverb⟩ → gracefully

where the symbol → means "can be rewritten as." Now, the structure above can be abstracted to define the formalized phrase-structure grammar.

Definition 2.1 A phrase-structure grammar G is a four-tuple $G = (V_N, V_T, P, S)$ in which (1) V_N and V_T are the nonterminal and terminal vocabularies (or variables) of G, respectively. In the example above,

$$V_N = \{⟨\text{sentence}⟩, ⟨\text{noun phrase}⟩, ⟨\text{verb phrase}⟩, ⟨\text{adjective}⟩$$
$$⟨\text{noun}⟩, ⟨\text{verb}⟩, ⟨\text{adverb}⟩\}$$

and

$$V_T = \{\text{the, girl, walks, gracefully}\}$$

The union of V_N and V_T constitutes the total vocabulary V of G, and $V_N \cap V_T$

$= \varnothing$. (2) P is a finite set of rewrite rules or productions denoted by $\alpha \to \beta$, where α and β are strings over V and with α involving at least one symbol of V_N. (3) $S \in V_N$ is the starting symbol of a sentence. This corresponds to the symbol "$\langle$sentence$\rangle$" in the foregoing example.

The following notations are frequently used.

(1) Σ^* is the set of all finite-length strings of symbols in a finite set of symbols Σ, including λ, the string of length 0. $\Sigma^+ = \Sigma^* - \{\lambda\}$.

(2) If x is a string, x^n is x written n times.

(3) $|x|$ is the length of the string x, or the number of symbols in string x.

(4) $\eta \underset{G}{\Longrightarrow} \gamma$, or a string η directly generates or derives another string γ if $\eta = \omega_1 \alpha \omega_2$, $\gamma = \omega_1 \beta \omega_2$, and $\alpha \to \beta$ is a member of P.

(5) $\eta \underset{G}{\overset{*}{\Longrightarrow}} \gamma$, or a string η generates or derives another string γ if there exists a sequence of strings ζ_1, ζ_2, $\ldots$, ζ_n such that $\eta = \zeta_1$, $\gamma = \zeta_n$, $\zeta_i \Longrightarrow \zeta_{i+1}$, $i = 1, 2, \ldots, n-1$. The sequence of strings ζ_1, ζ_2, $\ldots$, ζ_n is called a derivation of γ from η. It is clear that $\overset{*}{\Longrightarrow}$ is the reflexive and transitive closure of the relation $\Longrightarrow$.

The language generated by grammar G is

$$L(G) = \{x \,|\, x \in V_T^* \text{ such that } S \underset{G}{\overset{*}{\Longrightarrow}} x\} \qquad (2.1)$$

If G is a phrase-structure grammar, then $L(G)$ is called a phrase-structure language. A string of terminals and nonterminals γ is called a sentential form if $S \underset{G}{\overset{*}{\Longrightarrow}} \gamma$. Usually, if it is clear which grammar G is involved, the G under $\Longrightarrow$ or $\overset{*}{\Longrightarrow}$ can be omitted.

In natural languages, the problem of ambiguity often arises. One sentence may have several completely different meanings according to different ways of parsing it. For example, in the sentence, "They are rocking chairs," the word "rocking" can be interpreted either as part of the verb phrase "are rocking" or as part of the noun phrase modifying the noun "chairs." In the same manner, a formal grammar G is said to be ambiguous if there is a string $x \in L(G)$ that has more than one derivation. In pattern description languages, it is clear that ambiguity should be avoided; therefore, to find a family of unambiguous grammars is a problem of interest in this area.

Chomsky divided the phrase-structure grammars into four types according to the forms of the productions. In type 0 (unrestricted) grammars, there is no restriction on the productions, which may have any strings on either the right or the left of the substitution arrow. However, this type of grammar is too general to be useful, and the problem of whether a particular string is generated by a type 0 grammar is, in general, undecidable. The languages generated by type 0 grammars are called type 0 languages.

Type 1 (Context-Sensitive) Grammar

The productions of type 1 grammars are restricted to the form

$$\zeta_1 A \zeta_2 \rightarrow \zeta_1 \beta \zeta_2$$

where $A \in V_n$, ζ_1, ζ_2, $\beta \in V^*$, and $\beta \neq \lambda$. This can be read, "A can be replaced by β in the context ζ_1, ζ_2." It also implies that

$$|\zeta_1 A \zeta_2| \leq |\zeta_1 \beta \zeta_2| \qquad \text{or} \qquad |A| \leq |\beta|$$

The languages generated by context-sensitive grammars are called type 1, or context-sensitive, languages. An example of a context-sensitive language is

$$\{0^n 1^n 0^n \,|\, n = 1, 2, \ldots,\}$$

Type 2 (Context-Free) Grammar

The productions are of the form

$$A \rightarrow \beta$$

where $A \in V_N$ and $\beta \in V^+$. Note that a production of such a form allows the nonterminal A to be replaced by the string β independently of the context in which the A appears. The languages generated by context-free grammars are called type 2, or context-free, languages.

Example 2.1   Consider the grammar $G = (V_N, V_T, P, S)$, where $V_N = \{S, A, B\}$, $V_T = \{a, b\}$, and P:

(1)	$S \rightarrow aB$	(5)	$A \rightarrow a$
(2)	$S \rightarrow bA$	(6)	$B \rightarrow bS$
(3)	$A \rightarrow aS$	(7)	$B \rightarrow aBB$
(4)	$A \rightarrow bAA$	(8)	$B \rightarrow b$

The grammar G is context-free since for each production in P, the left part is a single nonterminal and the right part is a nonempty string of terminals and nonterminals. The language $L(G)$ is the set of all words in V_T^+ consisting of an equal number of a's and b's. Typical generations or derivations of sentences include

$$S \xoverset{(1)}{\Longrightarrow} aB \xoverset{(8)}{\Longrightarrow} ab$$

$$S \xoverset{(1)}{\Longrightarrow} aB \xoverset{(6)}{\Longrightarrow} abS \xoverset{(2)}{\Longrightarrow} abbA \xoverset{(5)}{\Longrightarrow} abba$$

$$S \xoverset{(2)}{\Longrightarrow} bA \xoverset{(5)}{\Longrightarrow} ba$$

$$S \xoverset{(2)}{\Longrightarrow} bA \xoverset{(4)}{\Longrightarrow} bbAA \xoverset{(5)}{\Longrightarrow} bbaA \xoverset{(5)}{\Longrightarrow} bbaa$$

where the parenthesized number indicates the production used.

An alternative method for describing any derivation in a context-free grammar is the use of derivation trees. A derivation tree for a context-free grammar $G = (V_N, V_T, P, S)$ can be constructed according to the following procedure.

(1) Every node of the tree has a label, which is a symbol of V.

(2) The root of the tree, that is, the node that no branch enters, has the label S.

(3) If a node has at least one descendant other than itself, and has the label A, then $A \in V_N$.

(4) If nodes $n_1, n_2, \ldots, n_k$ are the direct descendant of node n (with label A) in the order from the left, with labels $A_1, \ldots, A_k$, respectively, then

$$A \to A_1 A_2, \ldots, A_k$$

must be a production in P.

For example, the derivation $S \overset{*}{\Longrightarrow} abba$ in the grammar given in Example 2.1 can be described by the following derivation tree.

Similarly, the derivation tree for the sentence "The girl walks gracefully" is

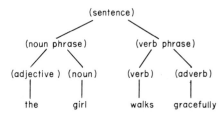

Type 3 (*Finite-State or Regular*) *Grammar*

The productions of type 3 grammars are of the form

$$A \to aB \quad \text{or} \quad A \to b$$

where $A, B \in V_N$ and $a, b \in V_T$. Note that A, B, a, b all are single symbols. For example, the language $\{0^m 1^n \mid m, n = 1, 2, \ldots\}$ is a finite-state language, while the language $\{0^n1^n \,|\, n = 1, 2, \ldots,\}$ is not. The languages generated by finite-state grammars are called type 3, or finite-state (or regular), languages.

Example 2.2   Consider the grammar $G = (V_N, V_T, P, S)$ where $V_N = \{S, A\}$, $V_T = \{a, b\}$, and P:

$$S \to aA$$
$$A \to aA$$
$$A \to b$$

Clearly, the grammar is a finite-state grammar. A typical sentence of $L(G)$ is generated by the derivation

$$S \Longrightarrow aA \Longrightarrow aaA \Longrightarrow aaaA \Longrightarrow aaab$$

In general, $L(G) = \{a^n b \,|\, n = 1, 2, \ldots\}$.

From the definitions of the four types of grammars, it should be clear that every finite-state grammar is context free; every context-free grammar is context sensitive; every context-sensitive grammar is type 0.

2.3   Finite-State Languages and Finite-State Automata

In Section 2.2, finite specifications for language from the viewpoint of generation (grammar) have been introduced. An alternative way of specifying a language is in terms of the set of strings that are accepted by a certain recognition device (recognition viewpoint). In this section the simplest recognizer, called a finite-state automaton, is introduced. The finite-state automaton can only accept all languages defined by the finite-state grammars. In later sections, recognizers for types 0, 1, and 2 languages will be introduced.

Definition 2.2   A (deterministic) finite-state automaton A is a quintuple

$$A = (\Sigma, Q, \delta, q_0, F)$$

where Σ is a finite set of input symbols (alphabet), Q is a finite set of states, δ is a mapping of $Q \times \Sigma$ into Q (next state function), $q_0 \in Q$ is the initial state, and $F \subseteq Q$ is the set of final states.

A convenient representation of a finite-state automaton is given in Fig. 2.1. The finite control, in one of the states in Q, reads symbols from an input tape in a sequential manner from left to right. Initially, the finite control is in state q_0 and is scanning the leftmost symbol of a string in Σ^*, which appears on the input tape. The interpretation of

$$\delta(q, a) = q', \quad q, q' \in Q \quad \text{and} \quad a \in \Sigma$$

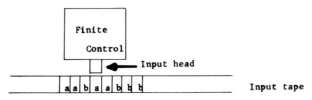

Fig. 2.1. A finite-state automaton.

is that the automaton A, in state q and scanning the input symbol a, goes to state q' and the input head moves one square to the right. A convenient way to represent the mapping δ is by the use of a state transition diagram. The state transition diagram corresponding to $\delta(q, a) = q'$ is shown in Fig. 2.2.

Fig. 2.2. Graphical representation of $\delta(q, a) = q'$.

The mapping δ can be extended from an input symbol to a string of input symbols by defining:

$$\delta(q, \lambda) = q, \qquad \delta(q, xa) = \delta(\delta(q, x), a), \qquad x \in \Sigma^* \quad \text{and} \quad a \in \Sigma$$

Thus, the interpretation of $\delta(q, x) = q'$ is that the automaton A, starting in state q, scanning through the string x on the input tape, will be in state q' and the input head moves right from the portion of the input tape containing x. A string or a sentence x is said to be accepted by A if

$$\delta(q_0, x) = p \qquad \text{for some} \quad p \in F$$

The set of strings accepted by A is defined as

$$T(A) = \{x \mid \delta(q_0, x) \in F\} \tag{2.2}$$

Example 2.3   Given a finite-state automaton $A = (\Sigma, Q, \delta, q_0, F)$, where $\Sigma = \{0, 1\}$, $Q = \{q_0, q_1, q_2, q_3\}$, and $F = \{q_0\}$. The state transition diagram of A is shown in Fig. 2.3. A typical sentence accepted by A is 101101, since

Fig. 2.3. State transition diagram of the automaton in Example 2.3.

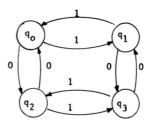

$\delta(q_0, 101101) = q_0 \in F$. In general, $T(A)$ is the set of all strings in $\{0, 1\}^*$ containing both an even number of 0's and an even number of 1's.

Definition 2.3 A nondeterministic finite-state automaton is a quintuple $(\Sigma, Q, \delta, q_0, F)$, where Σ is a finite set of input symbols (alphabet), Q is a finite set of states, δ is a mapping of $Q \times \Sigma$ into subsets of Q, $q_0 \in Q$ is the initial state, and $F \subseteq Q$ is the set of final states.

The only difference between the deterministic and nondeterministic case is that $\delta(q, a)$ is a (possibly empty) set of states rather than a single state. The interpretation of $\delta(q, a) = \{q_1, q_2, \ldots, q_l\}$ is that the automaton A, in state q, scanning a on its input tape, chooses any one of $q_1, \ldots, q_l$ as the next state and moves its input head one square to the right. As in the case of deterministic automata, the mapping δ can be extended from an input symbol to a string of input symbols by defining

$$\delta(q, \lambda) = \{q\}, \qquad \delta(q, xa) = \bigcup_{q_i \in \delta(q, x)} \delta(q_i, a), \qquad x \in \Sigma^* \quad \text{and} \quad a \in \Sigma$$

Furthermore, we can define

$$\delta(\{q_1, q_2, \ldots, q_l\}, x) = \bigcup_{i=1}^{l} \delta(q_i, x)$$

A string x is accepted by A if there is a state p such that

$$p \in \delta(q_0, x) \qquad \text{and} \qquad p \in F$$

The set of all strings accepted by A is defined as

$$T(A) = \{x \mid p \in \delta(q_0, x) \quad \text{and} \quad p \in F\} \tag{2.3}$$

Example 2.4 Given a nondeterministic finite-state automaton $A = (\Sigma, Q, \delta, q_0, F)$ where $\Sigma = \{0, 1\}$, $Q = \{q_0, q_1, q_2, q_3, q_4\}$, and $F = \{q_2, q_4\}$. The state transition diagram of A is shown in Fig. 2.4. A typical sentence accepted by A is 01011, since $\delta(q_0, 01011) = q_2 \in F$. In general, $T(A)$ is the set of all strings containing either two consecutive 0's or two consecutive 1's.

The relationship between deterministic finite-state automata and nondeterministic finite-state automata and the relationship between the finite-state languages and the sets of strings accepted by finite-state automata can be expressed by the following theorems.

Theorem 2.1 Let L be a set of strings accepted by a nondeterministic finite-state automaton $A = (\Sigma, Q, \delta, q_0, F)$. Then there exists a deterministic finite-state automaton $A' = (\Sigma', Q', \delta', q_0', F')$ that accepts L. The states of A' are all the subsets of Q, that is, $Q' = 2^Q$, and $\Sigma' = \Sigma$. F' is the set of all states in Q' containing a state of F. A state of A' will be denoted by $[q_1, q_2, \ldots, q_i] \in$

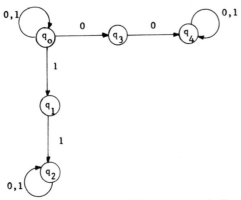

Fig. 2.4. State transition diagram of the automaton in Example 2.4.

Q' where $q_1, q_2, \ldots, q_i \in Q$. $q_0' = [q_0]$. $\delta'([q_1, \ldots, q_i], a) = [p_1, p_2, \ldots, p_j]$ if and only if $\delta(\{q_1, \ldots, q_i\}, a) = \bigcup_{k=1}^{i} \delta(q_k, a) = \{p_1, p_2, \ldots, p_j\}$.

Since the deterministic and nondeterministic finite-state automata accept the same sets of strings, we shall not distinguish between them unless it becomes necessary, but shall simply refer to both as finite-state automata.

Theorem 2.2   Let $G = (V_N, V_T, P, S)$ be a finite-state grammar. Then there exists a finite-state automaton $A = (\Sigma, Q, \delta, q_0, F)$ with $T(A) = L(G)$, where

(i) $\Sigma = V_T$;
(ii) $Q = V_N \cup \{T\}$;
(iii) $q_0 = S$;
(iv)   if P contains the production $S \to \lambda$, then $F = \{S, T\}$, otherwise, $F = \{T\}$;
(v)   the state T is in $\delta(B, a)$ if $B \to a$, $B \in V_N$, $a \in V_T$ is in P; and
(vi) $\delta(B, a)$ contains all $C \in V_N$ such that $B \to aC$ is in P and $\delta(T, a) = \varnothing$ for each $a \in V_T$.

Theorem 2.3   Given a finite-state automaton $A = (\Sigma, Q, \delta, q_0, F)$. Then there exists a finite-state grammar $G = (V_N, V_T, P, S)$ with $L(G) = T(A)$, where

(i) $V_N = Q$;
(ii) $V_T = \Sigma$;
(iii) $S = q_0$;
(iv) $B \to aC$ is in P if $\delta(B, a) = C$, $B, C \in Q$, $a \in \Sigma$; and
(v) $B \to a$ is in P if $\delta(B, a) = C$ and $C \in F$.

Example 2.5   Given the finite-state grammar $G = (V_N, V_T, P, S)$, where $V_N = \{S, B\}$, $V_T = \{a, b\}$, and P:

$$S \rightarrow aB$$
$$B \rightarrow aB$$
$$B \rightarrow bS$$
$$B \rightarrow a$$

A nondeterministic finite-state automaton $A = (\Sigma, Q, \delta, q_0, F)$ can be constructed such that $T(A) = L(G)$, where $\Sigma = V_T = \{a, b\}$, $Q = V_N \cup \{T\} = \{S, B, T\}$, $q_0 = S$, $F = \{T\}$, and δ is given by

$\delta(S, a) = \{B\}$ since $S \rightarrow aB$ is in P;
$\delta(S, b) = \varnothing$;
$\delta(B, a) = \{B, T\}$ since $B \rightarrow aB$ and $B \rightarrow a$ are in P;
$\delta(B, b) = \{S\}$ since $B \rightarrow bS$ is in P; and
$\delta(T, a) = \delta(T, b) = \varnothing$.

By Theorem 2.1 we can construct a deterministic finite-state automaton $A' = (\Sigma', Q', \delta', q_0', F')$ equivalent to A, where

$\Sigma' = \Sigma = \{a, b\}$, $Q' = \{\varnothing, [S], [B], [T], [S, B], [S, T], [B, T], [S, B, T]\}$
$q_0' = [S]$ $F' = \{[T], [S, T], [B, T], [S, B, T]\}$

$$\delta'([S], a) = [B], \qquad\qquad \delta'([S], b) = \varnothing$$
$$\delta'([B], a) = [B, T], \qquad\qquad \delta'([B], b) = [S]$$
$$\delta'([B, T], a) = [B, T], \qquad\qquad \delta'([B, T], b) = [S]$$
$$\delta'(\varnothing, a) = \delta'(\varnothing, b) = \varnothing$$

There are other transition rules of δ'. However, no states other than $\varnothing$, $[S]$, $[B]$, and $[B, T]$ will ever be entered by A', and the other states can be removed from Q' and F'.

2.4   Context-Free Languages and Pushdown Automata

In this section, we describe the two normal forms and some important properties of context-free grammars. The pushdown automaton is defined as the recognizer for context-free languages.

Theorem 2.4   (Chomsky normal form)   Any context-free language can be generated by a grammar $G = (V_N, V_T, P, S)$ in which all productions are of the form $A \rightarrow BC$ or $A \rightarrow a$ with $A, B, C \in V_N$, and $a \in V_T$.

Theorem 2.5 (Greibach normal form) Any context-free language can be generated by a grammar $G = (V_N, V_T P, S)$ in which every production is of the form $A \to a\alpha$ with $A \in V_N$, $a \in V_T$, and $\alpha \in V_N^*$.

Theorem 2.6 ("uvwxy theorem") Let L be any context-free language. There exist constants p and q depending only on L, such that if there is a string z in L, with $|z| > p$, then z may be written as $z = uvwxy$, where $|vwx| \le q$ and v and x are not both λ, such that for each integer $i \ge 0$, uv^iwx^iy is in L.

A context-free grammar G is said to be "self-embedding" if there is a nonterminal A with the property that $A \underset{G}{\overset{*}{\Longrightarrow}} \alpha_1 A \alpha_2$ where α_1, $\alpha_2 \in V^+$. The nonterminal A is also said to be self-embedding. It is noted that the self-embedding property gives rise to sentences of the form uv^iwx^iy and distinguishes a strictly context-free language from a finite-state language. If we have a context-free grammar G which is non-self-embedding, then we know that $L(G)$ is a finite-state language. Consequently, a context-free language is non-finite state (or nonregular) if and only if all of its grammars are self-embedding.

A pushdown automaton is essentially a finite-state automaton with an additional pushdown storage. The pushdown storage is a "first-in/last out" stack. That is, symbols may be entered (stored) or removed (read out) only at the top of the stack. The stack of trays on a spring that is often seen in cafeterias serves as a good example of pushown storage. The spring below the trays has just enough strength so that only one tray appears above the level of the counter. When the top tray is removed, the load on the spring is lightened, and the next tray in the stack appears above the level of the counter. If a tray is put on top of the stack, the pile is pushed down, and the tray just put on appears above the counter. For our purpose, the capacity of the stack can be arbitrarily large, so that we may add as many trays as we desire. A convenient representation of a pushdown automaton is shown in Fig. 2.5.

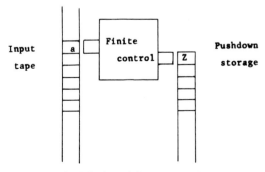

Fig. 2.5. A pushdown automaton.

Definition 2.4   Formally, a (nondeterministic) pushdown automaton M is defined as a 7-tuple $M = (\Sigma, Q, \Gamma, \delta, q_0, Z_0, F)$, where

Σ is a finite set of input symbols;
Q is a finite set of states;
Γ is a finite set of pushdown symbols;
$q_0 \in Q$ is the initial state;
$Z_0 \in \Gamma$ is the start symbol initially appearing on the pushdown storage;
$F \subseteq Q$ is the set of final states; and
δ is a mapping from $Q \times (\Sigma \cup \{\lambda\}) \times \Gamma$ to finite subsets of $Q \times \Gamma^*$.

The interpretation of

$$\delta(q, a, Z) = \{(q_1, \gamma_1), (q_2, \gamma_2), \ldots, (q_m, \gamma_m)\}$$

$$q, q_1, q_2, \ldots, q_m \in Q, \qquad a \in \Sigma, \qquad Z \in \Gamma, \qquad \gamma_1, \gamma_2, \ldots, \gamma_m \in \Gamma^*$$

$$(2.4)$$

is that the pushdown automaton in state q, with input symbol a and Z the top symbol on the pushdown storage, will, for any i, $1 \le i \le m$, enter state q_i, replace Z by γ_i, and advance the input head one symbol. When Z is replaced by γ_i, the leftmost symbol of γ_i will be placed highest on the pushdown storage and the rightmost symbol lowest. The interpretation of

$$\delta(q, \lambda, Z) = \{(q_1, \gamma_1), (q_2, \gamma_2), \ldots, (q_m, \gamma_m)\} \qquad (2.5)$$

is that the pushdown automaton in state q, independently of the input symbol being scanned and with Z the top symbol on the pushdown storage, will enter state q_i and replace Z by γ_i for any i, $1 \le i \le m$. In this case, the input head is not advanced. This is the operation used for manipulating the content of pushdown storage and is often called the "λ move."

In order to describe the "configuration" of a pushdown automaton, we use the ordered pair (q, γ), $q \in Q$ and $\gamma \in \Gamma^*$. For $a \in (\Sigma \cup \{\lambda\})$, γ, $\beta \in \Gamma^*$, and $Z \in \Gamma$, if $(q', \beta) \in \delta(q, a, Z)$, $q, q' \in Q$, the expression

$$a: (q, Z\gamma) \underset{M}{\longmapsto} (q', \beta\gamma)$$

means that according to the transition rules δ of the pushdown automaton, the input a may cause M to go from configuration $(q, Z\gamma)$ to configuration $(q', \beta\gamma)$. If for $a_1, a_2, \ldots, a_n$, $a_i \in (\Sigma \cup \{\lambda\})$, $1 \le i \le n$, states $q_1, q_2, \ldots, q_{n+1}$ and strings $\gamma_1, \gamma_2, \ldots, \gamma_{n+1}$, $\gamma_j \in \Gamma^*$, $1 \le j \le n+1$,

$$a_i: (q_i, \gamma_i) \underset{M}{\longmapsto} (q_{i+1}, \gamma_{i+1}), \qquad 1 \le i \le n$$

then we write

$$a_1 a_2 \cdots a_n: (q_1, \gamma_1) \underset{M}{\overset{*}{\longmapsto}} (q_{n+1}, \gamma_{n+1})$$

The M under $\longmapsto$ or $\overset{*}{\longmapsto}$ will be dropped whenever the meaning remains clear.

Two types of language acceptance are defined for pushdown automata. For a pushdown automaton M, the language accepted by final state $T(M)$ is defined as

$$T(M) = \{x \mid x\colon (q_0, Z_0) \overset{*}{\underset{M}{\longmapsto}} (q, \gamma) \quad \text{for any} \quad \gamma \in \Gamma^* \quad \text{and} \quad q \in F\} \qquad (2.6)$$

On the other hand, the language accepted by M by empty store $N(M)$ is defined to be

$$N(M) = \{x \mid x\colon (q_0, Z_0) \overset{*}{\underset{M}{\longmapsto}} (q, \lambda) \quad \text{for any} \quad q \in Q\} \qquad (2.7)$$

When accepting by empty store, the set of final states F is irrelevant. Thus, in this case we usually let $F = \varnothing$. The relationship between the languages accepted by pushdown automata by final state and those by empty store is expressed in the following theorem.

Theorem 2.7 L is $N(M_1)$ for some pushdown automaton M_1 if and only if L is $T(M_2)$ for some pushdown automaton M_2.

Example 2.6 The following pushdown automaton accepts $\{xcx^T \mid x \in \{0, 1\}^*\}$ by empty store:

$$M = (\Sigma, Q, \Gamma, \delta_0, q_0, Z_0, \varnothing)$$

where $\Sigma = \{0, 1, c\}$, $Q = \{q_1, q_2\}$, $\Gamma = \{R, B, G\}$, $q_0 = q_1$, $Z_0 = R$, and

$$
\begin{aligned}
&\delta(q_1, 0, R) = \{(q_1, BR)\}, && \delta(q_2, 0, B) = \{(q_2, \lambda)\} \\
&\delta(q_1, 0, B) = \{(q_1, BB)\}, && \delta(q_2, 1, G) = \{(q_2, \lambda)\} \\
&\delta(q_1, 0, G) = \{(q_1, BG)\}, && \delta(q_2, \lambda, R) = \{(q_2, \lambda)\} \\
&\delta(q_1, c, R) = \{(q_2, R)\}, && \delta(q_1, 1, R) = \{(q_1, GR)\} \\
&\delta(q_1, c, B) = \{(q_2, B)\}, && \delta(q_1, 1, B) = \{(q_1, GB)\} \\
&\delta(q_1, c, G) = \{(q_2, G)\} && \delta(q_1, 1, G) = \{(q_1, GG)\}
\end{aligned}
$$

Let $x = 001$; then $xcx^T = 011c100$. The sequence of transitions for the input sequence $001c100$ is

$$(q_1, R) \overset{0}{\longmapsto} (q_1, BR) \overset{0}{\longmapsto} (q_1, BBR) \overset{1}{\longmapsto} (q_1, GBBR)$$

$$\overset{c}{\longmapsto} (q_2, GBBR) \overset{1}{\longmapsto} (q_2, BBR) \overset{0}{\longmapsto} (q_2, BR)$$

$$\overset{0}{\longmapsto} (q_2, R) \overset{\lambda}{\longmapsto} (q_2, \lambda)$$

The following two theorems show that the languages accepted by nondeterministic pushdown automata by empty store are precisely the context-free languages.

Theorem 2.8   Let $L(G)$ be a context-free language generated by the grammar $G = (V_N, V_T, P, S)$ in Greibach normal form. Then there exists a nondeterministic pushdown automaton $M = (\Sigma, Q, \Gamma, \delta, q_0, Z_0, F)$ such that $N(M) = L(G)$, where

(i) $\Sigma = V_T$;
(ii) $Q = \{q_1\}$;
(iii) $\Gamma = V_N$;
(iv) $q_0 = q_1$;
(v) $Z_0 = S$;
(vi) $F = \varnothing$; and
(vii) $(q_1, \gamma) \in \delta(q_1, a, A)$ whenever $A \to a\gamma$ is in P.

Theorem 2.9   Let $N(M)$ be the language accepted by the pushdown automaton $M = (\Sigma, Q, \Gamma, \delta, q_0, Z_0, \varnothing)$. Then there exists a context-free grammar $G = (V_N, V_T, P, S)$ such that $L(G) = N(M)$, where

(i) $V_T = \Sigma$;
(ii) V_N is the set of triples of the form $[q, A, p]$, $q, p \in Q$ and $A \in \Gamma$, plus the new symbol S.
(iii) $S \to [q_0, Z_0, q]$ is in P for each $q \in Q$; and
(iv) $[q, A, p] \to a[q_1, B_1, q_2][q_2, B_2, q_3] \cdots [q_m, B_m, q_{m+1}]$ is in P for each $q, q_1, \ldots, q_{m+1} \in Q$ $(p = q_{m+1})$, each $a \in (\Sigma \cup \{\lambda\})$ and $A, B_1, \ldots, B_m \in \Gamma$, such that $(q_1, B_1 B_2 \cdots B_m) \cdots \in \delta(q, a, A)$. (If $m = 0$, then $q_1 = p$, $(p, \lambda) \in \delta(q, a, A)$ and $[q, A, p] \to a$ is in P.)

Example 2.7   Given a context-free grammar $G = (V_N, V_T, P, S)$ where $V_N = \{S, A, B\}$, $V_T = \{a, b\}$, and P:

$$S \to bA \qquad S \to aB$$
$$A \to a \qquad B \to b$$
$$A \to aS \qquad B \to bS$$
$$A \to bAA \qquad B \to aBB$$

A (nondeterministic) pushdown automaton $M = (\Sigma, Q, \Gamma, \delta, q_0, Z_0, \varnothing)$ can be constructed such that $N(M) = L(G)$, where $\Sigma = V_T = \{a, b\}$, $Q = \{q_1\}$, $\Gamma = V_N = \{S, A, B\}$, $q_0 = q_1$, and $Z_0 = S$; δ is given by:

$\delta(q_1, a, S) = \{(q_1, B)\}$ since $S \to aB$ is in P;
$\delta(q_1, b, S) = \{(q_1, A)\}$ since $S \to bA$ is in P;
$\delta(q_1, a, A) = \{(q_1, S), (q_1, \lambda)\}$ since $A \to aS$ and $A \to a$ are in P;
$\delta(q_1, b, A) = \{(q_1, AA)\}$ since $A \to bAA$ is in P;
$\delta(q_1, a, B) = \{(q_1, BB)\}$ since $B \to aBB$ is in P;
$\delta(q_1, b, B) = \{(q_1, S), (q_1, \lambda)\}$ since $B \to bS$ and $B \to b$ are in P.

By Theorem 2.8, $N(M) = L(G)$.

For finite-state automata, the deterministic and nondeterministic models are equivalent with respect to the languages accepted (Theorem 2.1). However, the same is not true for pushdown automata. For example, the language $L = \{xx^T \mid x \in \{0, 1\}^*\}$ is accepted by the pushdown automaton

$$M = (\Sigma, Q, \Gamma, \delta, q_0, Z_0, \varnothing)$$

where $\Sigma = \{0, 1\}$, $Q = \{q_1, q_2\}$, $\Gamma = \{A, B, C\}$, $q_0 = q_1$, $Z_0 = A$, and δ:

$\delta(q_1, 0, A) = \{(q_1, BA)\},$ $\delta(q_1, 1, C) = \{(q_1, CC), (q_2, \lambda)\}$

$\delta(q_1, 1, A) = \{(q_1, CA)\},$ $\delta(q_2, 0, B) = \{(q_2, \lambda)\}$

$\delta(q_1, 0, B) = \{(q_1, BB), (q_2, \lambda)\},$ $\delta(q_2, 1, C) = \{(q_2, \lambda)\}$

$\delta(q_1, 0, C) = \{(q_1, BC)\},$ $\delta(q_1, \lambda, A) = \{(q_2, \lambda)\}$

$\delta(q_1, 1, B) = \{(q_1, CB)\},$ $\delta(q_2, \lambda, A) = \{(q_2, \lambda)\}$

But L is not accepted by any deterministic pushdown automaton.

A language that is accepted by a deterministic pushdown automaton is called a deterministic context-free language. In Chapter 4, we shall define a subclass of the context-free grammars, called $LR(k)$ grammars, which generate deterministic context-free languages.

Definition 2.5   If every production of a context-free grammar is of the form $A \to uBv$ or $A \to u$, $A, B \in V_N$ and $u, v \in V_T^*$ and u and v are not both λ, then the grammar is called a linear grammar. Furthermore, the grammar is right-linear if $v = \lambda$ and left-linear if $u = \lambda$.

A language that can be generated by a linear grammar is called a linear language. Not all context-free languages are linear languages. Note that no intermediate string in a derivation of linear grammar contains more than one nonterminal, and a right-linear or left-linear grammar generates precisely finite-state languages.

2.5   Turing Machines and Linear-Bounded Automata

In this section we describe the third and the fourth types of recognizing devices and the languages accepted by them. They are the Turing machine and the linear-bounded automata. The basic model of a Turing machine consists of a finite control with a reading and writing head, and an input tape. The tape is infinite to the right. Initially, the n leftmost squares, for some finite n, hold the input string. The remaining infinity of squares hold the "blank," a special tape symbol B which is not an input symbol. (See Fig. 2.6). The difference between a Turing machine and a finite-state automaton is that the Turing machine, through its reading–writing head, can change symbols on the tape.

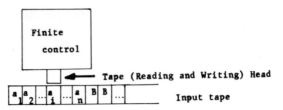

Fig. 2.6. A basic Turing machine.

Definition 2.6 A Turing machine T is a six-tuple $T = (\Sigma, Q, \Gamma, \delta, q_0, F)$ where Q is a finite set of states and Γ is a finite set of tape symbols. One of these is the blank symbol B. Σ, a subset of Γ not including B, is a set of input symbols; δ is a mapping from $Q \times \Gamma$ to $Q \times (\Gamma - \{B\}) \times \{L, R\}$ (it may be undefined for some arguments); $q_0 \in Q$ is the initial state; $F \subseteq Q$ is a set of final states.

A configuration of the Turing machine T is represented by the triple (q, α, i), where $q \in Q$, $\alpha \in (\Gamma - \{B\})^*$, the nonblank portion of the tape, and i is the distance of the tape (reading and writing) head of T from the left end of α. Let $(q, A_1 A_2 \cdots A_n, i)$ be a configuration of T, $1 \le i \le n + 1$. If $\delta(q, A_i) = (p, X, R)$, $1 \le i \le n$, then the operation of T, designated as an elementary move of T, can be expressed as

$$(q, A_1 A_2 \cdots A_n, i) \vdash_T (p, A_1 A_2 \cdots A_{i-1} X A_{i+1} \cdots A_n, i + 1)$$

That is, T writes the symbol X at the ith square of the tape and moves one square to the right. If $\delta(q, A_i) = (p, X, L)$, $2 \le i \le n$, then

$$(q, A_1 A_2 \cdots A_n, i) \vdash_T (p, A_1 A_2 \cdots A_{i-1} X A_{i+1} \cdots A_n, i - 1)$$

In this case, T writes X and moves left, but not off the left end of the tape. If $i = n + 1$, the tape head is reading the blank B. If $\delta(q, B) = (p, X, R)$, then

$$(q, A_1 A_2 \cdots A_n, n + 1) \vdash_T (p, A_1 A_2 \cdots A_n X, n + 2)$$

If, instead, $\delta(q, B) = (p, X, L)$, then

$$(q, A_1 A_2 \cdots A_n, n + 1) \vdash_T (p, A_1 A_2 \cdots A_n X, n)$$

If two configurations are related to each other by some finite number of elementary moves (including zero moves), the notation $\vdash_T^*$ will be used.

The language accepted by T is defined as

$$\{x \mid x \in \Sigma^* \text{ and } (q_0, x, 1) \vdash_T^* (q, \alpha, i) \text{ for some } q \in F, \alpha \in \Gamma^*, \text{ and } i\}$$

Given a T recognizing a language L, we assume that the T halts (i.e., that it has no next move) whenever the input is accepted. On the other hand, for strings not accepted, it is possible that T will not halt.

Example 2.8   The following Turing machine accepts $\{0^n 1^n \mid n \geq 1\}$.

$$T = (\Sigma, Q, \Gamma, \delta, q_0, F)$$

where $\Sigma = \{0, 1\}$, $Q = \{q_0, q_1, \ldots, q_5\}$, $\Gamma = \{0, 1, B, X, Y\}$, $F = \{q_5\}$, and δ:

$\delta(q_0, 0) = (q_1, X, R)$

$\delta(q_1, 0) = (q_1, 0, R)$, $\delta(q_1, Y) = (q_1, Y, R)$, $\delta(q_1, 1) = (q_2, Y, L)$

$\delta(q_2, Y) = (q_2, Y, L)$, $\delta(q_2, X) = (q_3, X, R)$, $\delta(q_2, 0) = (q_4, 0, L)$

$\delta(q_3, Y) = (q_3, Y, R)$, $\delta(q_3, B) = (q_5, Y, R)$

$\delta(q_4, 0) = (q_4, 0, L)$, $\delta(q_4, X) = (q_0, X, R)$

The sequence of operations for the input sequence $x = 0011$ is

$$(q_0, 0011, 1) \longmapsto (q_1, X011, 2) \longmapsto (q_1, X011, 3)$$
$$\longmapsto (q_2, X0Y1, 2) \longmapsto (q_4, X0Y1, 1)$$
$$\longmapsto (q_0, X0Y1, 2) \longmapsto (q_1, XXY1, 3)$$
$$\longmapsto (q_1, XXY1, 4) \longmapsto (q_2, XXYY, 3)$$
$$\longmapsto (q_2, XXYY, 2) \longmapsto (q_3, XXYY, 3)$$
$$\longmapsto (q_3, XXYY, 4) \longmapsto (q_3, XXYY, 5)$$
$$\longmapsto (q_5, XXYYY, 6), \qquad q_5 \in F$$

A nondeterministic Turing machine is a Turing machine of which the δ is a mapping from $Q \times \Gamma$ to subsets of $Q \times (\Gamma - \{B\}) \times \{L, R\}$. It can be shown that if a language L is accepted by a nondeterministic Turing machine T_1, then L is accepted by some deterministic Turing machine T_2.

Theorem 2.10   If L is generated by a type 0 grammar, then L is accepted by a Turing machine.

Theorem 2.11   If L is accepted by a Turing machine, then L is generated by a type 0 grammar.

Definition 2.7   A linear-bounded automaton M is a six-tuple $M = (\Sigma, Q, \Gamma, \delta, q_0, F)$ where

Q is a finite set of states;
Γ is a finite set of tape symbols;
$\Sigma \subseteq \Gamma$ is a set of input symbols;
δ is a mapping from $Q \times \Gamma$ to the subsets $Q \times \Gamma \times \{L, R\}$;
$q_0 \in Q$ is the initial state; and
$F \subseteq Q$ is a set of final states.

Σ contains two special symbols, usually denoted ¢ and \$, which are the left and right end marker, respectively, of the input string. Their function is to prevent the tape head from leaving the portion of tape upon which the input appears.

From Definition 2.7, a linear-bounded automaton is essentially a non-deterministic Turing machine that never leaves the portion of tape on which the input was placed. A configuration of M and the relation $\underset{M}{\longmapsto}$ between two configurations are defined as they were defined for the Turing machine. The language accepted by a linear-bounded automaton M is

$$\{x \mid x \in (\Sigma - \{¢, \$\})^* \text{ and } (q_0, ¢x\$, 1) \overset{*}{\underset{M}{\longmapsto}} (q, \alpha, i)$$
$$\text{for some } q \in F, \alpha \in \Gamma^* \text{ and } i\}$$

If the mapping $\delta(q, A)$, for any $q \in Q$ and $A \in \Gamma$, contains no more than one element, the linear-bounded automaton is deterministic. It is not yet known whether the class of languages accepted by nondeterministic linear-bounded automata properly contains that accepted by deterministic linear-bounded automata. It is true, of course, that any language accepted by a nondeter-ministic linear-bounded automaton is accepted by a deterministic Turing machine.

Theorem 2.12   If L is a context-sensitive language, then L is accepted by a (nondeterministic) linear-bounded automaton.

Theorem 2.13   If L is accepted by a linear-bounded automaton, then L is context sensitive.

A context-sensitive grammar $G = (V_N, V_T, P, S)$ is in Kuroda normal form [9] if each production in P is of one of the following forms:

$$A \to BB' \qquad AB \to A'B \qquad A \to a$$
$$AB \to AB' \qquad A \to A'$$

where $A, A', B, B' \in V_N$ and $a \in V_T$. Given any context-sensitive grammar G_1, we can effectively construct a context-sensitive grammar G_2 in Kuroda normal form such that $L(G_2) = L(G_1)$.

2.6   Modified Grammars

It is noted that context-free languages are not powerful enough to describe natural or programming languages. On the other hand, context-sensitive (but not context-free) languages are very complex for analysis. Consequently, a compromise is made by generalization of context-free grammars into programmed grammars and indexed grammars.

Programmed Grammars†

The special feature of a programmed grammar is that after applying a production to an intermediate string in a derivation, it is restricted as to which production may be applied next. Each production has a label; a core, consisting of a regular phrase-structure production; and two associated "go-to" fields. If possible, the production is applied to the intermediate string in a derivation and the next production to be used is selected from the first go-to field (success field). If the production cannot be applied to the string, then the next production is selected from the second go-to field (failure field).

Definition 2.8 A programmed grammar G_p is a five-tuple (V_N, V_T, J, P, S), where V_N, V_T, and P are finite sets of nonterminals, terminals, and productions, respectively. S is the starting symbol, $S \in V_N$. J is a set of production labels. A production in P is of the form:

$$(r) \qquad \alpha \to \beta \qquad S(U) \qquad F(W)$$

$\alpha \to \beta$ is called the core, where $\alpha \in V^* V_N V^*$ and $\beta \in V^*$. (r) is the label, $r \in J$. U is the success field and W the failure field. $U, W \subset J$.

The programmed grammar operates as follows. Production (1) is applied first. In general, if one tries to apply production (r) to rewrite a substring α and the current string η contains the substring α, then (r) $\alpha \to \beta$ is applied and the next production is selected from the success go-to field U. If the current string η does not contain α, then the production (r) is not used (i.e., η is not changed) and the next production is selected from the failure go-to field W. If the applicable go-to field contains $\varnothing$, the derivation halts. When the core of the production is of the context-free form (i.e., $A \to \beta$), the grammar is called a context-free programmed grammar. Similarly, if the core is of the finite-state or context-sensitive form, the grammar will be called a finite-state or a context-sensitive programmed grammar. It has been shown that the language generated by a context-free programmed grammar can be a context-sensitive language. Furthermore, it can be shown that the class of languages generated by a context-free programmed grammars properly includes the class of context-free languages.

Example 2.9 A programmed grammar G is constructed which generates context-sensitive language $\{0^n 1^n 0^n \mid n = 1, 2, \ldots,\}$.

† See Rosenkrantz [10].

Let $G = (\{S, A, B, C\}, \{0, 1\}, \{1, 2, \ldots, 8\}, P, S)$ and P consists of the following productions:

Label	Core	Success field	Failure field
1	$S \to ABC$	$S(\{2,5\})$	$F(\emptyset)$
2	$A \to 0A$	$S(\{3\})$	$F(\emptyset)$
3	$B \to 1B$	$S(\{4\})$	$F(\emptyset)$
4	$C \to 0C$	$S(\{2,5\})$	$F(\emptyset)$
5	$A \to 0$	$S(\{6\})$	$F(\emptyset)$
6	$B \to 1$	$S(\{7\})$	$F(\emptyset)$
7	$C \to 0$	$S(\{8\})$	$F(\emptyset)$
8	$S \to S$	$S(\emptyset)$	$F(\emptyset)$

It is seen that by using productions (1), (5), (6), (7), (8) sequentially, we obtain the following derivations:

$$S \Longrightarrow ABC \Longrightarrow 0BC \Longrightarrow 01C \Longrightarrow 010$$

If, on the other hand, starting from (1), the rules from production (2) to production (4) are applied n times, we obtain $0^{n-1}A1^{n-1}B0^{n-1}C$. Then, by applying productions (5), (6), (7), (8), we obtain $0^n1^n0^n$.

Indexed Grammars†

This generalization is accomplished by introducing another finite set, the set of flags to the grammar. Informally, the nonterminals in a string can be followed by arbitrary lists of flags. If a nonterminal, with flags following, is replaced by one or more nonterminals, the flags following each nonterminal are generated. If a nonterminal is replaced by terminals, the flags disappear.

Definition 2.9   An indexed grammar is a 5-tuple (V_N, V_T, F, P, S) where V_N, V_T, F, and P are finite sets of nonterminals, terminals, flags, and productions, respectively. $S \in V_N$ is the starting symbol. The flags in F are finite-sets of index productions of the form $A \to \alpha$, where $A \in V_N$ and $\alpha \in V^*$. The productions in P are of the form $A \to X_1 \psi_1 X_2 \psi_2, \ldots, X_m \psi_m$ where $X_1, \ldots, X_m$ are in V_N or V_T and $\psi_1, \psi_2, \ldots, \psi_m$ in F^*. If X_i is in V_T, then $\psi_i = \lambda$.

Define the relation $\Longrightarrow$ on strings in $(V_N F^* \cup V_T)^*$ as follows:

1. If $\alpha A \theta \beta$ is a string, with α and $\beta \in (V_N F^* \cup V_T)^*$, $A \in V_N$, $\theta \in F^*$, and $A \to X_1 \psi_1 X_2 \psi_2, \ldots, X_m \psi_m$ is in P, with $X_i \in V$ and $\psi_i \in F^*$ for all i, then we

† See Aho [11].

may write

$$\alpha A \theta \beta \Longrightarrow \alpha X_1 \varphi_1 X_2 \varphi_2, \ldots, X_m \varphi_m \beta$$

Here, φ_i is λ if $X_i \in V_T$ and $\varphi_i = \psi_i \theta$ if $X_i \in V_N$.

2. If $\alpha A f \theta \beta$ is a string with α and $\beta \in (V_N F^* \cup V_N)^*$, $A \in V_N$, $f \in F$, and $\theta \in F^*$, and $A \to X_1 X_2, \ldots, X_m$ is an index production in f, then we may write

$$\alpha A f \theta \beta \Longrightarrow \alpha X_1 \varphi_1 X_2 \varphi_2, \ldots, X_m \varphi_m \beta$$

where φ_i is λ if $X_i \in V_T$ and $\varphi_i = \theta$ if $X_i \in V_N$.

It has been shown that the class of indexed languages properly includes the class of context-free languages and is a proper subset of the class of context-sensitive languages. Furthermore, indexed grammars are sufficiently powerful to specify many of the non-context-free features of the algorithmic programming languages.

Example 2.10   An indexed grammar G is constructed which generates context-sensitive language $\{0^n 1^n 0^n \mid n = 1, 2, \ldots\}$.

Let $G = (\{S, T, A, B\}, \{0, 1\}, \{f, g\}, P, S)$ where P consists of the following productions:

(1) $S \to Tg$
(2) $T \to Tf$
(3) $T \to ABA$

with

(4) $f = \{A \to 0A, B \to 1B\}$
(5) $g = \{A \to 0, B \to 1\}$

Applying production (1) once and production (2) $(n - 1)$ times, and finally (3), we have

$$S \Longrightarrow Tg \Longrightarrow Tfg \Longrightarrow, \ldots, \Longrightarrow Tf^{n-1}g \Longrightarrow Af^{n-1}gBf^{n-1}gAf^{n-1}g$$

Then, using the flags (4) and (5), we have

$$Af^{n-1}gBf^{n-1}gAf^{n-1}g \Longrightarrow 0Af^{n-2}g1Bf^{n-2}g0Af^{n-2}g \Longrightarrow, \ldots,$$
$$\Longrightarrow 0^{n-1}Ag1^{n-1}Bg0^{n-1}Ag \Longrightarrow 0^n 1^n 0^n$$

References

1. N. Chomsky, Three models for the description of language. *IEEE Trans. Information Theory* **IT-2**, 113–124 (1956).
2. N. Chomsky, On certain formal properties of grammars. *Information and Control* **2**, 137–167 (1959).

3. N. Chomsky, A note on phrase-structure grammars. *Information and Control* **2**, 393–395 (1959).
4. N. Chomsky and G. A. Miller, Finite state languages. *Information and Control* **1**, 91–112 (1958).
5. Y. Bar-Hillel, C. Gaifman, and E. Shamir, On categorial and phrase-structure grammars *Bull. Res. Counc. Isr. Sect. 9F*, 1–16 (1960).
6. Y. Bar Hillel, M. Perles, and E. Shamir, On formal properties of simple phrase Structure grammars. *Z. Phonetik, Sprachwiss. Kommunikat.* **14**, 143–172 (1961).
7. A. V. Aho and J. D. Ullman, The theory of languages. *J. Math. Syst. Theory* **2**, 97–125 (1968).
8. J. E. Hopcroft and J. D. Ullman, *Formal Languages and Their Relation to Automata.* Addison-Wesley, Reading, Massachusetts, 1969.
9. S. Y. Kuroda, Classes of languages and linear-bounded Automata. *Information and Control* **7**, 207–223 (1964).
10. D. J. Rosenkrantz, Programmed grammars; A new device for generating formal languages. *IEEE Annu. Symp. Switching and Automata Theory, 8th, Austin, Texas, 1967,* Conf. Record.
11. A. V. Aho, Indexed grammars: An extension of context free grammars. *IEEE Annu. Symp. Switching and Automata Theory, 8th, Austin, Texas, 1967,* Conf. Record.
12. T. L. Booth, *Sequential Machines and Automata Theory.* Wiley, New York, 1967.
13. S. A. Greibach, A new normal form theorem for context-free phrase structure grammars. *J. Assoc. Comput. Mach.* **12**, 42–52 (1965).
14. N. Chomsky, Context-free grammars and pushdown storage. Quart. Progr. Dept. No. 65. Res. Lab. Elect., MIT, Cambridge, Massachusetts, 1962.
15. M. O. Rabin and D. Scott, Finite automata and their decision problems. *IBM J. Res. Develop.* **3**, 114–125 (1959).
16. R. J. Nelson, *Introduction to Automata.* Wiley, New York, 1968.
17. M. L. Minsky, *Computation: Finite and Infinite Machines.* Prentice-Hall, Englewood Cliffs, New Jersey, 1967.
18. M. A. Harrison, On the relation between grammars and automata. *Advances in Information Syst. Sci.* **4**, 39–92 (1972).
19. S. Ginsburg, *The Mathematical Theory of Context-Free Languages.* McGraw-Hill, New York, 1966.
20. T. Kasai, An hierarchy between context-free and context-sensitive languages. *J. Comput. System Sci.* **4**, 492–508 (1970).

Chapter 3

Languages for Pattern Description

3.1 Selection of Pattern Primitives

As was discussed in Section 1.1, the first step in formulating a syntactic model for pattern description is the determination of a set of primitives in terms of which the patterns of interest may be described. This determination will be largely influenced by the nature of the data, the specific application in question, and the technology available for implementing the system. There is no general solution for the primitive selection problem at this time. The following requirements usually serve as a guide for selecting pattern primitives.

(i) The primitives should serve as basic pattern elements to provide a compact but adequate description of the data in terms of the specified structural relations (e.g., the concatenation relation).

(ii) The primitives should be easily extracted or recognized by existing nonsyntactic methods, since they are considered to be simple and compact patterns and their structural information not important.

For speech patterns, phonemes are naturally considered as a " good " set of primitives with the concatenation relation.† Similarly, strokes have been

† The view of continuous speech as composed of one sound segment for each successive phoneme is, of course, a simplification of facts.

47

suggested as primitives in describing handwriting. However, for general pictorial patterns, there is no such "universal picture element" analogous to phonemes in speech or strokes in handwriting. Sometimes, in order to provide an adequate description of the patterns, the primitives should contain the information that is important to the specific application in question. For example, if size (or shape or location) is important in the recognition problem, then the primitives should contain information relating to size (or shape or location), so that patterns from different classes are distinguishable by whatever method is to be applied to analyze the descriptions. This requirement often results in a need for semantic information in describing primitives, which will be discussed in Section 3.4. The following simple example is used to illustrate that, for the same data, different problem specifications would result in different selections of primitives.

Example 3.1   Suppose that the problem is to discriminate rectangles (of different sizes) from nonrectangles. The following set of primitives is selected:

$$
\begin{array}{lll}
a' & 0° & \text{horizontal line segment} \\
b' & 90° & \text{vertical line segment} \\
c' & 180° & \text{horizontal line segment} \\
d' & 270° & \text{vertical line segment}
\end{array}
$$

The set of all rectangles (of different sizes) is represented by a single sentence or string $a'b'c'd'$.

If the problem is also to discriminate rectangles of different sizes, the foregoing description would be inadequate. An alternative is to use unit-length line segments as primitives:

The set of rectangles of different sizes can then be described by the language

$$L = \{a^n b^m c^n d^m \mid n, m = 1, 2, \ldots\} \tag{3.1}$$

Requirement (ii) may sometimes conflict with requirement (i) because the primitives selected according to requirement (i) may not be easy to recognize using existing techniques. On the other hand, requirement (ii) could allow the selection of quite complex primitives as long as they can be recognized.

With more complex primitives, simpler structural descriptions (e.g., simple grammars) of the patterns could be used. This trade-off may become quite important in the implementation of the recognition system. An example is the recognition of two-dimensional mathematical expressions in which characters and mathematical notations are primitives (Appendix C). However, if we consider the characters as subpatterns and describe them in terms of simpler primitives (e.g., strokes or line segments), the structural descriptions of mathematical expressions would be more complex than in the case where characters are used directly as primitives.

Another example is the recognition of Chinese characters (Appendix E) [1–3]. From knowledge about the structure of Chinese characters, a small number of simple segmentation operations, such as

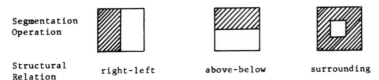

Segmentation Operation

Structural Relation   right-left   above-below   surrounding

can be used. Each operation also generates a particular structural relation between the two neighboring subpatterns or primitives. Applying these operations recursively, that is, segmentizing each subpattern again by any one of the three operations, we can segmentize a Chinese character into its subpatterns and primitives. If the primitives can be extracted or recognized by existing techniques, a Chinese character can be described syntactically with the given set of structural relations. It is anticipated that the resulting structural descriptions will be much more complex if we choose basic strokes as the primitives.

One of the earliest papers describing the decomposition of pictorial patterns into primitives [4] presented a conceptually appealing method which allows the recognition system to (heuristically) determine the primitives by inspection of training samples. A pattern is first examined by a programmed scan. The result of the scan is to produce descriptions of segments of the picture (subpictures) that are divisions conveniently produced by the scanning process, and not necessarily true divisions. The scanning process also includes preprocessing routines for noise cleaning, gap filling, and curve following. The subpictures obtained in the scan are analyzed and connected, when appropriate, into true picture parts; a description is given in terms of the length and slope of straight-line segments and the length and curvature of curved segments. The structural relations among various segments (primitives) of a picture are expressed in terms of a connection table (table of joins). The assembly program produces a "statement" which gives a complete description of the pattern. The description is independent of the orientation and the size of the picture, the lengths of the various parts being given relative to one another. It is, in effect, a coded representation of the pattern and may be regarded as a one-dimensional string consisting of symbols chosen from a specified alphabet. The coded representation gives the length, slope, and curvature of each primitive, together with details of the ends and joins to other primitives. No explicit consideration is given to formalizing the pattern syntax.

In the recognition of the program, a comparison or match is made between the statement or coded string describing the unknown pattern and the statements along with the (class) name of the patterns they describe already stored within the computer. To reduce the search time, the stored statements are examined and arranged into groups according to some common feature. For example, all statements describing pictures with, say, four ends would be grouped together. Then, at an early stage in the recognition process, a statement referring to an unknown pattern that has four ends would be compared with the statements that are members of the "four ends" group. The criterion for a successful match is that there must be a certain measure of agreement between the unknown statement and the statement referring to the pattern with which it is to be identified. This agreement must be substantially greater than the agreement between the unknown statement and any of the other stored statements. The program uses a sequence of tests, from the comparison of a small number of important features to that of the detailed statement. At each stage, the statement or statements which give a high degree of agreement will be given the next stage test against the unknown statement. The recognition scheme, though heuristically developed, was considered quite successful, allowing for such annoyances as varied pattern size, pattern rotation, etc., but it required an impractical amount of processing time and a large dictionary in order to store the valid pattern structures.

A formal model for the abstract description of English cursive script has been proposed by Eden and Halle [5]. The primitives are four distinct line segments in the form of a triple:

$$\sigma_j = [(x_{j_1}, y_{j_1}), (x_{j_2}, y_{j_2}), \theta_j] \qquad\qquad (3.2)$$

where (x_j, y_j)'s represent the approximate location of the end points of the line segment, and θ_j refers to the sense of rotation from the first to the second end point. θ_j is positive if the sense of rotation is clockwise and negative if it is counterclockwise. The four primitives are

$$\sigma_1 = [(1, 0), (0, 0), +] \qquad\qquad \text{"bar"}$$

$$\sigma_2 = [(1, 1), (0, 0), +] \qquad\qquad \text{"hook"}$$

$$\sigma_3 = [(0, 0), (0, 1), +] \qquad\qquad \text{"arch"}$$

$$\sigma_4 = [(1, \varepsilon), (0, 0), +], \qquad 0 < \varepsilon < 1, \qquad \text{"loop"}$$

They can be transformed by changing the sign of θ or by reflection about the horizontal or vertical axis. These transformations generate 28 strokes (because of symmetry, the arch generates only 4 strokes), but only 9 of them are of interest in the English script commonly used.

A word is completely specified by the stroke sequence constituting its letters. A word is represented by the image of a mapping of a finite sequence of strokes into the set of continuous functions, the mapping being specified by concatenation and tracing rules applied in specific order. Only two concatenation rules are required. The first specifies stroke locations within a letter. The rule prescribes that two consecutive strokes are concatenated by identifying the abscissa of the terminal end point of the first stroke with that of the initial end point of the second stroke. The second rule states that across a letter boundary, the leftmost end point of the stroke following the boundary is placed so as to be to the right of the rightmost end point of the penultimate stroke before the letter boundary. The simple cursive strokes of the word "globe" are shown in Fig. 3.1 and their concatenation in Fig. 3.2. These

Fig. 3.1. Cursive strokes of the word "globe."

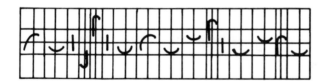

Fig. 3.2. Stroke sequence representation of the word "globe."

concatenation rules are not sufficient to specify all sequences of English letters unambiguously. Nevertheless, the ambiguities are intrinsic to the writing system, even in careful handwriting.

No formal syntax was attempted for the description of handwriting. Interesting experimental results on the recognition of cursive writing were obtained by Earnest [6] and Mermelstein [7] using a dictionary and rather heuristic recognition criteria. In addition, the dynamics of the trajectory (in space and time) that the point of the pen traces out as it moves across the paper has also been studied [8]. The motion of the pen is assumed to be controlled by a pair of orthogonal forces, as if one pair of muscles controls the vertical displacement and another the horizontal.

More general methods for primitive selection may be grouped roughly into methods emphasizing boundaries and methods emphasizing regions. These methods are discussed in the following sections.

3.1.1 Primitive Selection Emphasizing Boundaries or Skeletons

A set of primitives commonly used to describe boundaries or skeletons is the chain code devised by Freeman [9, 10]. Under this scheme, a rectangular grid is overlaid on the two-dimensional pattern, and straight-line segments are used to connect the grid points falling closest to the pattern. Each line segment is assigned an octal digit, according to its slope. The pattern is thus represented by a chain (or string) or chains of octal digits. Figure 3.3 illustrates

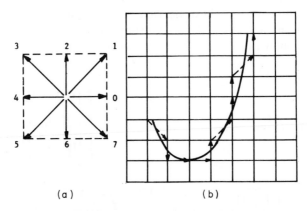

(a)   (b)

Fig. 3.3. Freeman's chain code. (a) Octal primitives; (b) coded string of the curve = 7600212212.

the primitives and the coded string describing a curve. This descriptive scheme has some useful properties. For example, patterns coded in this way can be rotated through multiples of 45° simply by adding an octal digit (modulo 8) to every digit in the string (however, only rotations by multiples of 90° can be accomplished without some distortion of the pattern). Other simple manipulations, such as expansion, measurement of curve length, and determination of pattern self-intersections, are easily carried out. Any desired degree of resolution can be obtained by adjusting the fineness of the grid imposed on the patterns. This method is, of course, not limited to simply connected closed boundaries; it can be used for describing arbitrary two-dimensional figures composed of straight or curved lines and line segments.

Notable work using Freeman's chain code includes efforts by Knoke and Wiley [11] and Feder [12]. Knoke and Wiley attempted to demonstrate that linguistic approaches can usually be applied to describe structural relationships within patterns (hand-printed characters, in this case). The following example is used to illustrate their procedure. After preprocessing of a pattern, say the character A, to obtain its skeleton, a chain code is used to initially describe the pattern. For the character A, the initial description using the chain code is shown in Fig. 3.4. Each line segment is represented by a chain-

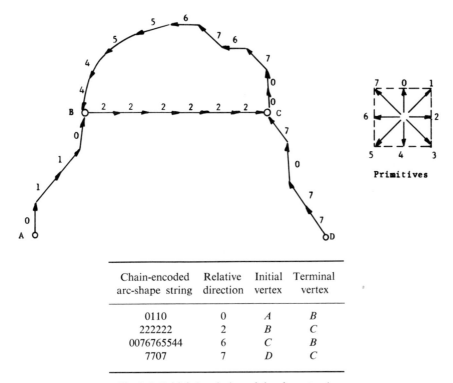

Chain-encoded arc-shape string	Relative direction	Initial vertex	Terminal vertex
0110	0	A	B
222222	2	B	C
0076765544	6	C	B
7707	7	D	C

Fig. 3.4. Initial description of the character A.

encoded arc-shape string, a relative direction, and its initial and terminal vertices. A "transformational grammar" [13, 14] is then applied to reduce irregularities (i.e., to smooth the pattern). The rules of transformation are given in Fig. 3.5. After the arc-shape labels are substituted for the chain-encoded string in the description, each pattern may be represented by a finite, planar, directed, connected graph, and the structure of any such graph can be completely described in terms of its arcs and vertices. The list of vertex interconnections and arc-shape labels constitutes a sentence in the pattern language. For character A, the resultant description after the transformation is shown in the following tabulation.

	Arc-shape string	Arc-shape label	Relative direction	Initial vertex	Terminal vertex
	0110		0	A	B
	222222		2	B	C
	0076765544		6	C	B
	7707		7	D	C
		LIN	0	A	B
L_s		LIN	2	B	C
transformation		CURL	6	C	B
$\longrightarrow$		LIN	7	D	C

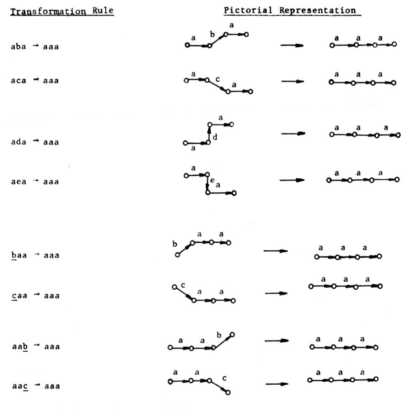

Transformation Rule   **Pictorial Representation**

aba → aaa

aca → aaa

ada → aaa

aea → aaa

b̲aa → aaa

c̲aa → aaa

aab̲ → aaa

aac̲ → aaa

Fig. 3.5a. L_s transformation rules (arc-shape language), where $a \in \{0, 1, 2, 3, 4, 5, 6, 7\}$, $b = a + 1$ (modulo 8), $c = a - 1$ (modulo 8), $d = a + 2$ (modulo 8), $e = a - 2$ (modulo 8), g means that a is the first or last element in a string.

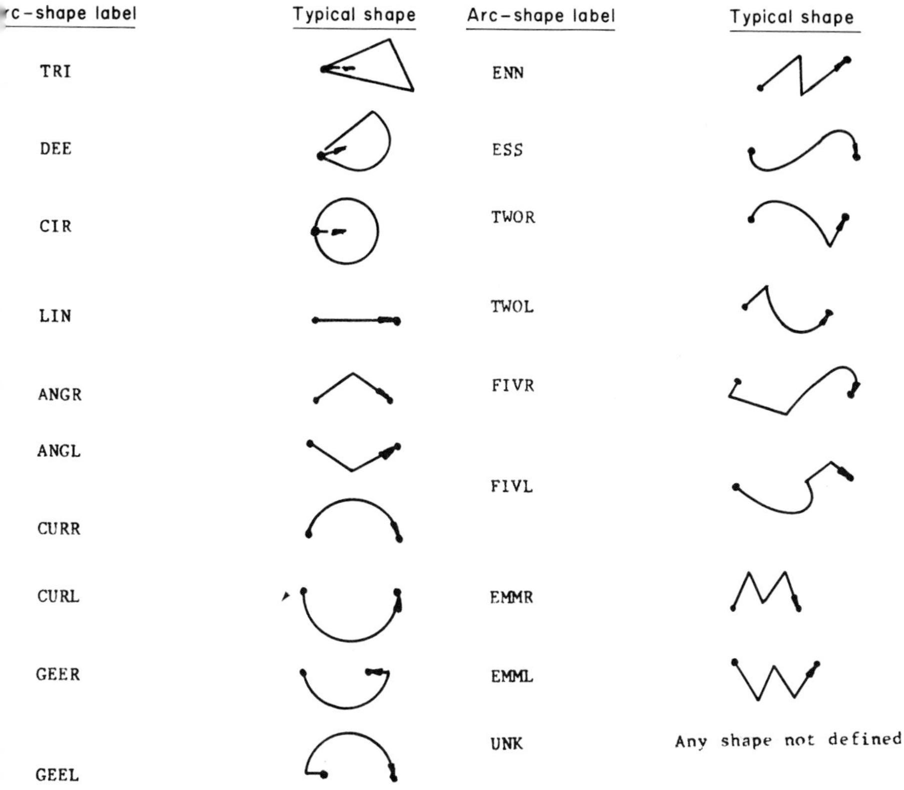

rc-shape label	Typical shape	Arc-shape label	Typical shape
TRI		ENN	
DEE		ESS	
CIR		TWOR	
LIN		TWOL	
ANGR		FIVR	
ANGL			
		FIVL	
CURR			
CURL		EMMR	
GEER		EMML	
		UNK	Any shape not defined
GEEL			

Fig. 3.5b. Arc-shape labels (arrowhead indicates the relative direction).

Since the number of arc-shape labels and the number of vertices are finite, the character description language L_c is a finite-state language. However, Knoke and Wiley used the following context-free language L_c'.

$$L_c': \quad V_N = \{S, H, T, R\}, \qquad V_T = \{l_1, \ldots, l_{152}, A, B, C, D, E, F, G\}$$

and P:

$$S \rightarrow H$$
$$S \rightarrow SH$$
$$H \rightarrow TRT$$
$$T \rightarrow A, T \rightarrow B, T \rightarrow C, T \rightarrow D, T \rightarrow E, T \rightarrow F, T \rightarrow G$$
$$R \rightarrow l_1, R \rightarrow l_2, \ldots, R \rightarrow l_{152}$$

where H is the phrase, T the vertex, and R the arc-shape relative-direction pair. Since there are 19 different arc-shape labels and 8 relative directions, there are $19 \times 8 = 152$ arc-shape relative-direction pairs. It is clear that $L_c \subset L_c'$.

Recognition is performed by matching L_c character descriptions to a dictionary of L_c prototypes. The matching algorithm consists of four steps.

Step 1. Obtain standard form description for the "unknown" input pattern.

Step 2. Property screening—Check the number of vertices and the arc-shape labels as a preliminary screening.

Step 3. Structure matching ignoring relative directions—Create all possible relabelings of prototype vertices to attempt to obtain a restricted (ignoring relative directions) one-to-one match between each prototype description and the "unknown" description.

Step 4. Complete structure matching—A candidate prototype which results from the Step 3 matching must be checked exhaustively for relative direction matches. If a match is found, recognition has been completed. If not, the algorithm returns to Step 3 unless all possible vertex relabelings have been generated and checked.

Although some limited success has been achieved in character recognition, this method has some difficulties, namely,

(1)   it has not been possible to find a pattern grammar L_c which generates all and only valid patterns;

(2)   it is necessary to restrict the allowable length of pattern sentences;

(3)   it has not been possible to utilize context;

(4)   the syntax analysis (matching algorithm) is slow and inefficient; and

(5)   only connected patterns are considered.

Nevertheless, the idea of using a local smoothing transformation on the chain-coded strings and relabeling them as new primitives for pattern description has since been used by a number of other authors [15, 16]. The L_s transformational grammar transforms the original octal primitives into a set of new primitives (arc-shape labels) to make the description of patterns more efficient (shorter sentences in this case).

Feder's work considers only patterns that can be encoded as strings of primitives. Several bases for developing pattern languages are discussed, including equations in two variables (straight lines, circles and circular arcs, etc.), pattern properties (self-intersections, convexity, etc.), and various measures of curve similarity. The computational power (automaton complexity) required to detect the elements of these pattern languages is studied. However, this problem is complicated considerably by the fact that (i) these languages are mostly context sensitive and not context free, (ii) the chain code yields only a piecewise linear approximation of the original pattern, and (iii) the coding of a typical curve is not unique, depending to a degree on its location and orientation with respect to the coding grid.

Other applications of the chain code include the description of contour maps [17], "shape matching" [18], and the identification of high-energy particle tracks in bubble chamber photographs [19]. Contour lines can be encoded as chains. Contour map problems may involve finding the terrain to be flooded by a dam placed at a particular location, the watershed area for a river basin, or the terrain visible from a particular mountain-top location; or the determination of optimum highway routes through mountainous terrain. In shape matching, two or more two-dimensional objects having irregular contours are to be matched for all or part of their exterior boundary. For some such problems the relative orientation and scale of the objects to be matched may be known and only translation is required. The problem of matching aerial photographs to each other as well as to terrain maps falls into this category. For other problems, either orientation, or scale, or both may be unknown and may have to be determined as part of the problem. An example of problems in which relative orientation has to be determined is that of the computer assembly of potsherds and jigsaw puzzles [20].

Other syntactic pattern recognition systems using primitives with the emphasis on boundary, skeleton, or contour information include systems for hand-printed character recognition [21–23], bubble chamber and spark chamber photograph classification [24–26], chromosome analysis [27, 28], fingerprint identification [29–32], face recognition [33, 34], and scene analysis [35–37].

3.1.2 Pattern Primitives in Terms of Regions

A set of primitives for encoding geometric patterns in terms of regions has been proposed by Pavlidis [38]. In this case, the basic primitives are half planes in the pattern space† (or the field of observation). It can be shown that any figure can be approximated by an arbitrary polygon which in turn may be expressed as the union of a finite number of convex polygons. Each convex polygon can, in turn, be represented as the intersection of a finite number of half planes. By defining a suitable ordering (a sequence) of the convex polygons composing the arbitrary polygon, it is possible to determine a unique minimal set of maximal (in an appropriate sense) polygons, called primary subsets, the union of which is the given polygon. In linguistic analogy, a figure can be thought of as a "sentence," the convex polygons composing it as "words," and the half planes as "letters." This process is summarized in this section.

† This could be generalized to half spaces of the pattern space.

Let A be a bounded polygon and let $s_1, s_2, \ldots, s_n$ be its sides. The following definition is introduced.

Definition 3.1 A point x in the plane will be said to be positive with respect to one side if it lies on the same side of the extension of a side as the polygon does with respect to the side itself. Otherwise, it will be said to be negative with respect to that side.

Example 3.2 For the polygon A given in Fig. 3.6, the point x is positive with respect to the sides s_5 and s_6, but negative with respect to s_7. Similarly, y is positive with respect to s_4 and s_7, but negative with respect to s_5. If we extend all the sides of A in both directions, A is intersected by some of these extensions, and it is subdivided into $A_1, A_2, \ldots, A_9$ convex polygons.

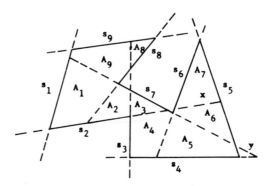

Fig. 3.6. Polygon A of Example 3.2.

Obviously, the points that are positive with respect to a side form a half plane whose boundary is the extension of the side. Let h_i denote the half plane corresponding to the side s_i, and let Q denote the intersection of all the half planes $h_1, h_2, \ldots, h_n$ in A. If A is convex, then $A = Q$. If A is not convex, then Q may be empty or simply different from A. Let Q_I represent the intersection of all the half planes except $s_{i_1}, \ldots, s_{i_k}$ where $I = \{i_1, \ldots, i_k\}$, the index set. Then we can define a sequence of Q_I as follows:

$$Q = \bigcap_{i=1}^{n} h_i, \quad Q_j = \bigcap_{\substack{i=1 \\ i \neq j}}^{n} h_i, \quad Q_{jk} = \bigcap_{\substack{i=1 \\ i \neq j, i \neq k}}^{n} h_i, \ldots \qquad (3.3)$$

This is an increasing sequence, since $Q \subset Q_j \subset Q_{jk} \ldots$ The last element of the sequence will be the whole plane, and it is obtained for $I = \{1, \ldots, n\}$. If a sequence of the preceding form has a maximal element, in the sense that any subsequent members are not subsets of A, then that set is called a primary (convex) subset of *A*. A nonempty member of such a *Q* sequence that is also a subset of *A* is called a nucleus of *A* if all the previous elements of the sequence are empty. Consequently, it can be shown that the union of the primary subsets of *A* precisely equals *A*.

For a given polygon the primary subsets can be found by forming all the sequences Q, Q_j, Q_{jk}, ... and searching for their maximal elements. This is a well-defined procedure and, hence, the primary subsets of *A* are unique. An example is given here to illustrate this procedure. For the polygon shown in Fig. 3.7, the intersection of all the half planes *Q* is the triangle *CPR*. It is

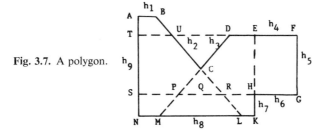

Fig. 3.7. A polygon.

easily seen that this is also the intersection of the half planes h_2, h_3, and h_6; therefore, only the sets Q_2, Q_3, and Q_6 will be different from Q. The set Q_2 is the trapezoid *PDEH*, the set Q_3 the trapezoid *TURS*, and the set Q_6 the triangle *MCL*. For each one of the sets, the same procedure can be applied. The sets Q_{23} (rectangle *TEHS*) and Q_{24} are not included in the polygon, and hence are not considered any further. On the other hand, the set Q_{26} is the trapezoid *DEKM*, which is the intersection of the half planes h_3, h_4, h_7, and h_8. If any one of the four half planes is removed, the resulting sets are not subsets of the polygon, and consequently, Q_{26} is a primary subset. The procedure can be represented by a treelike diagram such as that shown in Fig. 3.8. The first number on each node represents the half plane missing (in addition to the previous ones) and the second number (with parentheses) denotes the indices of the minimum number of half planes necessary for the intersection. In this example, in addition to Q_{26}, sets Q_{27} (trapezoid *DFGP*) and Q_{364} (trapezoid *ABLN*) are the only other primary subsets.

If the original figure (polygon) has not been approximated by an arbitrary polygon but by one whose sides are parallel to certain prespecified directions $\varphi_1, \varphi_2, \ldots, \varphi_k$, then the use of finite directions allows the representation of a polygon by half planes from a finite alphabet, rather than by a finite number of convex polygons. Any one of the half planes determined by the sides of the polygon will actually be a parallel translation of the half planes H_i determined by the chosen directions. These half planes play the role of primitives. If *G*

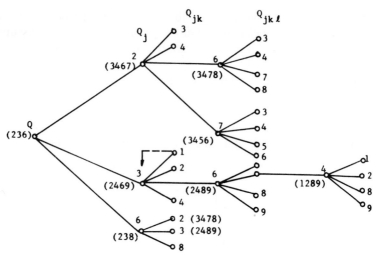

Fig. 3.8. Primary subset diagram for the polygon in Fig. 3.7.

denotes the group of parallel translations, then each primary subset R_j can be expressed as

$$R_j = \bigcap_{i=1}^{2k} g_i{}^j H_i, \qquad g_i{}^j \in G \tag{3.4}$$

where the intersection goes over $2k$ because each direction defines two half planes. If a half plane does not actually appear in the formation of R_j, it will still be included in the expression above with an arbitrary transformation g which will be required only to place its boundary outside the field of observation. The original polygon A can then be represented as

$$A = \bigcup_{j=1}^{l} R_j = \bigcup_{j=1}^{l} \bigcap_{i=1}^{2k} g_i{}^j H_i \tag{3.5}$$

If a similarity measure between two convex polygons A and B, denoted by $S(A, B)$, can be appropriately defined, we may be able to find a finite set B of convex polygons $B_1, \ldots, B_N$ such that for every convex polygon A of interest there will exist a member of B and a member of G to satisfy

$$S(A, gB) < \delta$$

for a prespecified δ. The members of B will be referred to as the basic components, and consequently we can write

$$A = \bigcup_{i=1}^{l} g_i B_{k(i)} \tag{3.6}$$

An appropriate similarity measure should be selected in accordance with our intuitive notion of similarity, and it should also be invariant under similarity transformations, that is, transformations which do not change the shape of the figures. If G is a group of such transformations, then an equivalence relation between the figures A and B can be defined as

$$A \sim B \quad \text{if there exists a} \quad g \in G \quad \text{such that} \quad A = gB$$

This equivalence relation induces a partition of the set of figures into disjoint classes. If d is a metric defined on the set of figures, then the "distance" (not necessarily a metric) on the set of equivalence classes can be defined as

$$D([A], [B]) = \min_{g, f \in G} d(gA, fB) \tag{3.7}$$

For bounded convex polygons, a good candidate for such a "distance" is the one induced by the Hansdorff metric [39]. Let G be a group of isomorphisms (e.g., parallel translations); then (3.7) can be simplified to

$$H([A], [B]) = \min_{g \in G} h(A, gB) = \min_{g \in G} h(gA, B) \tag{3.8}$$

where h is the Hansdorff distance. It can easily be shown that in this case H is actually a metric.

Another similarity measure that can be defined for convex polygons with sides parallel to prespecified directions is based on the fact that such a polygon can be represented by a sequence of numbers denoting the length of the corresponding sides. For example, if the directions are $0°$, $45°$, $90°$, and $135°$, the polygon A in Fig. 3.9 will be represented by $l_1, l_2, l_3, l_4, l_5, l_6, l_7, l_8$ if the

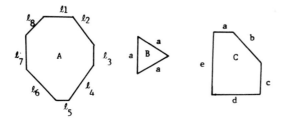

Fig. 3.9. Convex polygons.

starting point is taken to be the one of the leftmost side parallel to $0°$. Similarly, the triangle B will be represented by $0, a, 0, a, 0, 0, a, 0$ and the polygon C by $a, b, c, 0, d, 0, e, 0$. This allows us to define a similarity measure as

$$D(A, B) = \left[\sum_{i=1}^{8} (l_i^A - l_i^B)^2 \right]^{1/2} \tag{3.9}$$

which is invariant under translation and is obviously a metric. If invariance under scaling is desired, the definition above can be modified to

$$D'(A, B) = \underset{m}{\text{Min}} \left[\sum_{i=1}^{8} (l_i^A - ml_i^B)^2 \right]^{1/2} \qquad (3.10)$$

If a convex polygon has a side parallel to none of the given directions, it can be approximated along different directions. (For example, if the side has an angle with the horizontal of 23°, it could be approximated by either 0° or 45°.)

After a finite set of basic components and a (finite) set of transformations are determined, we may introduce a binary matrix whose columns correspond to the basic components $B_1, \ldots, B_N$ and the rows to the elements of G. An element of the binary matrix is equal to 1 if the corresponding combination of g_i and $B_{k(i)}$ is present in (3.6) and 0 otherwise. Then the recognition will be based on the presence of certain members of the basic components and on their interrelation. This binary matrix can also be reduced to a binary vector by a concatenation of its columns or rows, and then many of the decision-theoretic classification techniques can be applied.

It is noted that this approach provides a formalism for describing the syntax of polygonal figures and of more general figures which can be approximated reasonably well by polygonal figures. The analysis or recognition procedure requires the definition of suitable measures of similarity between polygons. The similarity measures considered so far are quite sensitive to noise in the patterns and/or are difficult to implement practically on a digital computer. A somewhat more general selection procedure of pattern primitives based on regions has recently been proposed by Rosenfeld and Strong [40].

Another way of representing polygonal figures is by the use of primary graphs [41, 42]. The primary graph of a polygon A is one whose nodes correspond to the nuclei and the primary subsets of A, and whose branches connect each nucleus to all the primary subsets containing it. An example is given in Fig. 3.10. Primary subsets and nuclei of polygons approximating the figures are shown in Fig. 3.10a. (Shaded areas are nuclei.) Primary graphs for the corresponding polygons in (a) are given in Fig. 3.10b. This form of representation may not characterize a figure uniquely; however, it does provide information describing it, and in particular, about its topology. Also, as will be seen later in this chapter, patterns represented by graphs can be formally described by graph grammars.

Another approach to the analysis of geometric patterns using regions is discussed primarily in relation to the problem of scene analysis [36, 43]. Minsky and Papert [44] have considered the direct transformation of a gray-scale picture to regions, bypassing the edge-finding, line-fitting procedures.

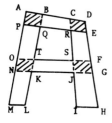
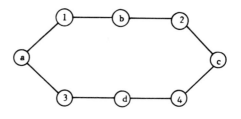

Primary subset or nucleus	Label
ABLM	a
ADEP	b
CDHI	c
OFGN	d
ABQP	1
CDER	2
OTKN	3
SFGJ	4

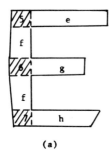
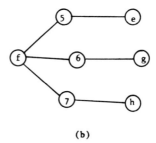

(a)   (b)

Fig. 3.10. (a) Polygonal figures and (b) corresponding primary graphs

Regions are constructed as the union of squares whose corners have the same or nearly the same gray scale. The method proposed by Guzman [45] assumes that a picture can be reduced by preprocessing to a list of vertices, lines, and surfaces. Various heuristics, based on the analysis of types of intersections of lines and surfaces, are applied to this list to compose its elements into two- or three-dimensional regions. Some candidate pattern recognition schemes have been investigated, all of which involve methods for matching the reduced pattern descriptions against a prototype dictionary. The procedure studied by Brice and Fennema [46] decomposes a picture into atomic regions of uniform gray scale. A pair of heuristics is used to join these regions in such a way as to obtain regions whose boundaries are determined more by the natural lines of the scene than by the artificial ones introduced by quantization and noise. Then a simple line-fitting technique is used to approximate the region boundaries by straight lines, and finally the scene analyzer interprets the picture using some simple tests on object groups generated by a Guzman-like procedure.

3.2   Pattern Grammar: Introduction

Assume that a satisfactory solution of the primitive selection problem is available for a given application. The next step is the construction of a grammar (or grammars) which will generate a language (or languages) to describe the patterns under study. Ideally, it would be nice to have a grammatical inference machine which would infer a grammar from a set of given strings describing the patterns under study. Unfortunately, such a machine has not been available except for some very special cases (see Chapter 7). In most cases so far, the designer constructs the grammar based on the a priori knowledge available and his experience. It is known that increased descriptive power of a language is paid for in terms of increased complexity of the analysis system (recognizer or acceptor). Finite-state automata are capable of recognizing or accepting finite-state languages, although the descriptive power of finite-state languages is also known to be weaker than that of context-free and context-sensitive languages. On the other hand, nonfinite, nondeterministic devices are required, in general, to accept the languages generated by context-free and context-sensitive grammars. Except for the class of deterministic languages,† nondeterministic parsing procedures are usually needed for the analysis of context-free languages. The trade-off between the descriptive power and the analysis efficiency of a grammar for a given application is, at present, almost completely justified by the designer. (For example, a precedence language may be used for pattern description in order to obtain good analysis efficiency; or, on the other hand, a context-free programmed grammar generating a context-sensitive language may be selected in order to describe the patterns effectively.) The effect of the theoretical difficulty may not be serious, in practice, as long as some care is exercised in developing the required grammars. This is especially true when the languages of interest are actually finite-state, even though the form of the grammars may be context-sensitive, or when the languages may be approximated by finite-state languages. The following simple examples illustrate some of the points just discussed, particularly the increased power of the productions of the more general classes of grammars.

† For example, see Sections 4.4 and 4.6 for $LR(k)$ languages and precedence languages.

Example 3.3   It is desired to construct a grammar to generate the finite-state language $L = \{a^n b^n c^n \mid 1 \leq n \leq 3\}$. This might be the language describing, say, the set of equilateral triangles of side length one, two, or three units. In order for the grammar to be compatible with a top-down goal-oriented analysis procedure, the grammar must produce terminals in a strictly left-to-right order, and at most one terminal may be produced by a single application of any production. Nonterminals may not appear to the left of terminal symbols, but the generation of nonterminals is otherwise unrestricted.

(1)   A finite-state grammar $G_1 = (V_N, V_T, P, S)$ where

$$V_N = \{S, A_1, A_2, B_{10}, B_{20}, B_{30}, B_{21}, B_{31}, B_{32}, C_1, C_2, C_3\} \qquad V_T = \{a, b, c\}$$

and P:

$$
\begin{array}{lll}
S \rightarrow aA_1 & B_{10} \rightarrow bC_1 & B_{32} \rightarrow bC_3 \\
S \rightarrow aB_{10} & B_{20} \rightarrow bB_{21} & C_1 \rightarrow c \\
A_1 \rightarrow aA_2 & B_{21} \rightarrow bC_2 & C_2 \rightarrow cC_1 \\
A_1 \rightarrow aB_{20} & B_{30} \rightarrow bB_{31} & C_3 \rightarrow cC_2 \\
A_2 \rightarrow aB_{30} & B_{31} \rightarrow bB_{32} &
\end{array}
$$

(2)   A context-free grammar (in Greibach normal form) $G_2 = (V_N, V_T, P, S)$ where

$$V_N = \{S, A_1, A_2, B_1, B_2, B_3, C\}, \qquad V_T = \{a, b, c\}$$

and P:

$$
\begin{array}{lll}
S \rightarrow aA_1 C & A_1 \rightarrow aA_2 C & B_2 \rightarrow bB_1 \\
A_1 \rightarrow b & A_2 \rightarrow aB_3 C & B_1 \rightarrow b \\
A_1 \rightarrow aB_2 C & B_3 \rightarrow bB_2 & C \rightarrow c
\end{array}
$$

(3)   A context-free programmed grammar $G_3 = (V_N, V_T, J, P, S)$ where

$$V_N = \{S, B, C\}, \qquad V_T = \{a, b, c\}, \qquad J = \{1, 2, 3, 4, 5\}$$

and P:

Label	Core	Success Field	Failure Field
1	$S \rightarrow aB$	$\{2, 3\}$	$\{\varnothing\}$
2	$B \rightarrow aBB$	$\{2, 3\}$	$\{\varnothing\}$
3	$B \rightarrow C$	$\{4\}$	$\{5\}$
4	$C \rightarrow bC$	$\{3\}$	$\{\varnothing\}$
5	$C \rightarrow c$	$\{5\}$	$\{\varnothing\}$

Even for this simple case, the context-free grammar is considerably more compact than the finite-state grammar. For this example, a context-sensitive grammar would not be much different from the context-free grammar and, hence, has not been given. However, the context-free programmed grammar is still more compact than the context-free grammar.

Example 3.4   The language $L = \{a^n b^n c^n d^n \mid n \geq 1\}$ could be interpreted as the language describing squares of side length $n = 1, 2, \ldots$:

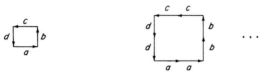

L is known as a context-sensitive language, and can be generated in the following two ways:

(1)   A context-sensitive grammar $G_1 = (V_N, V_T, P, S)$ where

$$V_N = \{S, A, B, C, D, E, F, G\}, \qquad V_T = \{a, b, c, d\}$$

and P:

$$
\begin{array}{lll}
S \rightarrow aAB & DB \rightarrow FB & dFB \rightarrow dFd \\
A \rightarrow aAC & Ed \rightarrow Gd & dFd \rightarrow Fdd \\
A \rightarrow D & cG \rightarrow Gc & cF \rightarrow Fc \\
Dc \rightarrow cD & dG \rightarrow Gd & bF \rightarrow bbc \\
Dd \rightarrow dD & aG \rightarrow abcD & aF \rightarrow ab \\
DC \rightarrow EC & bG \rightarrow bbcD & bB \rightarrow bcd \\
EC \rightarrow Ed & &
\end{array}
$$

(2)   A context-free programmed grammar $G_2 = (V_N, V_T, P, S, J)$ where

$$V_N = \{S, A, B, C, D\}, \qquad V_T = \{a, b, c, d\}, \qquad J = \{1, 2, 3, 4, 5, 6, 7\}$$

and P:

Label	Core	Success Field	Failure Field
1	$S \rightarrow aAB$	$\{2, 3\}$	$\{\emptyset\}$
2	$A \rightarrow aAC$	$\{2, 3\}$	$\{\emptyset\}$
3	$A \rightarrow D$	$\{4\}$	$\{\emptyset\}$
4	$C \rightarrow d$	$\{5\}$	$\{6\}$
5	$D \rightarrow bDc$	$\{4\}$	$\{\emptyset\}$
6	$B \rightarrow d$	$\{7\}$	$\{\emptyset\}$
7	$D \rightarrow bc$	$\{\emptyset\}$	$\{\emptyset\}$

It is noted that if n is finite, then $L = \{a^n b^n c^n d^n \,|\, n = 1, \ldots, N\}$ can certainly also be generated by a finite-state or context-free grammar.

It should be remarked that a grammar is most appropriate for description when the pattern of interest is built up from a small set of primitives by recursive application of a small set of production rules. Also, primitive selection and grammar construction should probably be treated simultaneously rather than in two different stages. There is no doubt that a different selection of pattern primitives will result in a different grammar for the description of a given set of patterns. Sometimes, a compromise is necessary in order to develop a suitable grammar. Example 3.1 may also be used to illustrate this point. Referring to the example, it is evident that the grammar which generates $L = \{a^n b^m c^n d^m \,|\, n, m = 1, 2, \ldots\}$ will be much more complex than the grammar generating $a'b'c'd'$.

Although many classes of patterns appear to be intuitively context sensitive, context-sensitive (but not context-free) grammars have rarely been used for pattern description simply because of their complexity. Context-free languages have been used to describe patterns such as Latin characters (Appendixes B and D), chromosome images (Appendix A), spark chamber pictures (Appendix B), chemical structures (Appendix G), and spoken digits (Appendix F).

In addition to (i) the trade-off between the language descriptive power and the analysis efficiency, and (ii) the compromise sometimes necessary between the primitives selected and the grammar constructed, the designer should also be aware of the need to control the excessive strings generated by the constructed grammar. The number of pattern strings available in practice is always limited. However, in most cases, the grammar constructed would generate a large or infinite number of strings.† It is hoped that the excessive strings generated are similar to the available pattern strings. Unfortunately, this may not be true, since the grammar, in many cases, is constructed heuristically. The problem may become very serious when the excessive strings include some pattern strings that should belong to other classes. In this case, adjustments should be made to exclude these strings from the language generated by the constructed grammar.

Recently, probably due to their relative effectiveness in describing natural-language structures, transformational grammars have been proposed for pattern description [47, 48]. Transformational grammars would allow the

---

† It may be argued that, in practice, a pattern grammar can always be finite state, since it is constructed from a finite number of pattern strings. However, the finite-state grammar so constructed may require a large number of productions. In such a case, a context-free or a context-free programmed pattern grammar may be constructed for the purpose of significantly reducing the number of productions.

possibility of determining from the pattern generative mechanism a simple base grammar (deep structure) which generates a certain set of patterns and a problem-oriented set of transformations. Through the base grammar and the transformations, the original set of patterns can be described.

From the foregoing discussion, it might be concluded that, before efficient grammatical inference procedures are available, a man–machine interactive system would be suitable for the problem of grammar construction. The basic grammar and the various trade-offs and compromises have to be determined by the designer. The results of any adjustment on the grammar constructed can be easily checked and displayed through a computer system. As will be seen later in Chapter 7, even in the development of grammatical inference procedures, man–machine interactive systems are often suggested to make the proposed inference procedure more efficient and computationally manageable.

3.3   High Dimensional Pattern Grammars

3.3.1   General Discussion

In describing patterns using a string grammar, the only relation between subpatterns and/or primitives is the concatenation; that is, each subpattern or primitive can be connected only at the left or right. This one-dimensional relation has not been very effective in describing two- or three-dimensional patterns. A natural generalization is to use a more general formalism, including other useful relations [49–55]. Let R be a set of n-ary relations $(n \geq 1)$. A relation $r \in R$ satisfied by the subpatterns and/or primitives $X_1, \ldots,$ X_n is denoted $r(X_1, \ldots, X_n)$. For example, TRIANGLE (a, b, c) means that the ternary relation TRIANGLE is satisfied by the line segments a, b, and c, and ABOVE (X, Y) means that X is above Y. The following examples illustrate pattern descriptions using this formalism of relations.

Example 3.5   The mathematical expression

$$\frac{a + b}{c}$$

can be described by

$$\text{ABOVE}(\text{ABOVE}(\text{LEFT}(a, \text{LEFT}(+, b)), \text{—}), c)$$

where LEFT(X, Y) means that X is to the left of Y.

Example 3.6   The following grammar will generate sentences describing houses [55a].

$$G = (V_N, V_T, P, S)$$

where

$$V_N = \{\langle \text{house}\rangle, \langle \text{side view}\rangle, \langle \text{front view}\rangle, \langle \text{roof}\rangle, \langle \text{gable}\rangle,$$
$$\langle \text{wall}\rangle, \langle \text{chimney}\rangle, \langle \text{windows}\rangle, \langle \text{door}\rangle\}$$

$$V_T = \left\{ \; ☐, \; ◖, \; ▣, \; ⊞, \; △, \; □, \; ▱, \; →, \; (\cdot), \; ⊙, \; ◡, \; ↑, \; ↦ \; \right\}$$

$S = \langle \text{house}\rangle$

P: $\langle \text{door}\rangle \rightarrow$ ▣

$\langle \text{windows}\rangle \rightarrow$ ⊞ , $\langle \text{windows}\rangle \rightarrow \rightarrow (\langle \text{windows}\rangle,$ ⊞ $)$

$\langle \text{chimney}\rangle \rightarrow$ ☐ , $\langle \text{chimney}\rangle \rightarrow$ ◖

$\langle \text{wall}\rangle \rightarrow$ □ , $\langle \text{wall}\rangle \rightarrow$ ◡ $(\langle \text{door}\rangle,$ □ $)$

$\langle \text{wall}\rangle \rightarrow$ ⊙ $(\langle \text{windows}\rangle,$ □ $)$

$\langle \text{gable}\rangle \rightarrow$ △ , $\langle \text{gable}\rangle \rightarrow$ ↑ $(\langle \text{chimney}\rangle,$ △ $)$

$\langle \text{roof}\rangle \rightarrow$ ▱ , $\langle \text{roof}\rangle \rightarrow$ ↑ $(\langle \text{chimney}\rangle,$ ▱ $)$

$\langle \text{front view}\rangle \rightarrow$ ↑ $(\langle \text{gable}\rangle, \langle \text{wall}\rangle)$

$\langle \text{side view}\rangle \rightarrow$ ↑ $(\langle \text{roof}\rangle, \langle \text{wall}\rangle)$

$\langle \text{house}\rangle \rightarrow \langle \text{front view}\rangle$

$\langle \text{house}\rangle \rightarrow$ ↦ $(\langle \text{house}\rangle, \langle \text{side view}\rangle)$

The notation

→   (X, Y)   means that X is to the right of Y;

⊙   (X, Y)   means that X is inside of Y;

◡   (X, Y)   means that X is inside on the bottom of Y;

↑   (X, Y)   means that X rests on top of Y;

↦   (X, Y)   means that X rests to the right of Y.

House	Description
$\uparrow$	$\uparrow\,(\,\triangle\,,\,\square\,)$
	$\uparrow\,(\,\uparrow\,(\,\triangledown\,,\,\triangle\,),\,\bigcirc\,(\,\square\,,\,\square\,)\,)$
	$\mapsto(\,\uparrow\,(\,\uparrow\,(\,\square\,,\,\triangledown\,),\,\odot\,(\,\boxplus\,,\,\square\,)),\,\uparrow\,(\,\triangle\,,\,\bigcirc\,(\,\square\,,\,\square\,))\,)$

A simple two-dimensional generalization of string grammars is to extend grammars for one-dimensional strings to two-dimensional arrays [56, 57]. The primitives are the array elements and the relation between primitives is the two-dimensional concatenation. Each production rewrites one subarray by another, rather than one substring by another. Relationships between array grammars and array automata (automata with two-dimensional tapes) have been studied recently [58]. A typical example is given in the following example [47, 56].

Example 3.7 The following grammar generates all possible 45° right-angled triangles:

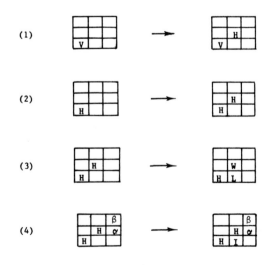

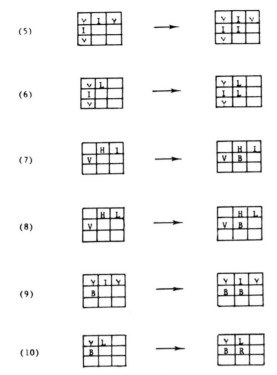

where $\alpha = L$ or I (interior point); $\beta = H$ (hypoteneuse point) or W; $\gamma = $ anything, including blank.

Initial symbol S =

A typical generation of a triangle is

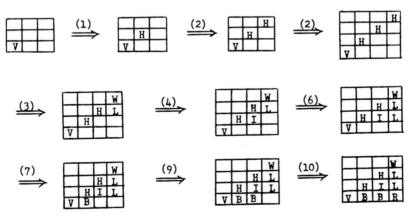

3.3.2   Special Grammars

Shaw, by attaching a "head" (hd) and a "tail" (tl) to each primitive, has used the four binary operators $+$, $\times$, $-$, and $*$ for defining binary concatenation relations between primitives (Appendix B).

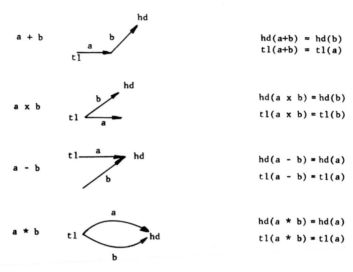

a + b		$hd(a+b) = hd(b)$ $tl(a+b) = tl(a)$
a x b		$hd(a \times b) = hd(b)$ $tl(a \times b) = tl(b)$
a - b		$hd(a - b) = hd(a)$ $tl(a - b) = tl(a)$
a * b		$hd(a * b) = hd(a)$ $tl(a * b) = tl(a)$

For string languages, only the operator $+$ is used. In addition, the unary operator $\sim$ acting as a tail/head reverser is also defined; that is,

$$hd(\sim a) = tl(a)$$
$$tl(\sim a) = hd(a)$$

In the case of describing patterns consisting of disconnected subpatterns, the "blank" or "don't care" primitive is introduced. Each pictorial pattern is represented by a "labeled branch-oriented graph" where branches represent primitives. A context-free string grammar is used to generate the Picture Description Language (PDL) (Appendix B).

Once we include relations other than just concatenations, the description of a pattern may be more conveniently represented by a relational graph, where nodes represent subpatterns or primitives and branches denote (binary) relations.† (Refer to Section 1.1.) When n-ary relations are involved, a graph-representable description can be obtained by transforming all relations into binary ones. A unary relation $r(X)$ can be changed to a binary one $r'(X, \lambda)$

---

† This straightforward representation is called a "labeled node-oriented directed graph." See Appendix E and Clowes [47] for examples using a tree-representable description.

where λ denotes the "null" primitive. The relation $r(X_1, \ldots, X_n)$ $(n > 2)$ can be transformed into a composition of binary relations, such as

$$r_1(X_1, r_2(X_2, \ldots, r_{n-1}(X_{n-1}, X_n)))$$

or into a conjunction of binary relations

$$r_1(X_{11}, X_{12}) \wedge r_2(X_{21}, X_{22}) \wedge \cdots \wedge r_k(X_{k1}, X_{k2})$$

or into a combination of these. For example, the ternary relation TRIANGLE (a, b, c) could be transformed into either one of the following equivalent binary relations:

$$\text{CAT}(a, b) \wedge \text{CAT}(b, c) \wedge \text{CAT}(c, a) \qquad \text{or} \qquad \Delta(b, \text{CAT}(a, c))$$

where $\text{CAT}(X, Y)$ means that $\text{hd}(X)$ is concatenated to $\text{tl}(Y)$, that is, $\text{CAT}(X, Y) = X + Y$, and $\Delta(X, Y)$ means that the line X is connected to form a triangle with the object Y consisting of two concatenated segments. In general, replacement of an n-ary relation with binary ones using composition requires the addition of more levels in the description.

Based on an idea in the work of Narasimhan [21], Feder has formalized a "plex" grammar which generates languages with terminals having an arbitrary number of attaching points for connecting to other primitives or sub-patterns (Appendix G). The primitives of the plex grammar are called N-attaching point entities (NAPEs). Each production of the plex grammar is in a context-free form in which the connectivity of primitives or subpatterns is described by using explicit lists of labeled concatenation points (called joint lists). While the sentences generated by a plex grammar are not directed graphs, they can be transformed either by assigning labeled nodes to both primitives and concatenation points, as suggested by Pfaltz and Rosenfeld [59] or by transforming primitives to nodes and concatenations to labeled branches. Figure 3.11 gives an example illustrating such transformations [60].

Pfaltz and Rosenfeld have extended the concept of string grammars to grammars for labeled graphs called webs (Appendix H). Labeled node-oriented graphs are explicitly used in the productions. Each production describes the rewriting of graph α into another graph β and also contains an "embedding" rule E which specifies the connection of α to its surrounding graph (host web) when α is rewritten. The formalism of web grammars has been rather extensively analyzed [59]. The relations among PDL grammars, plex grammars, and web grammars have been discussed by Shaw [60] and Rosenfeld [61].

Pavlidis [62, 63] has generalized string grammars to graph grammars by including nonterminal symbols which are not simple branches or nodes. An mth-order nonterminal structure is defined as an entity that is connected to the rest of the graph by m nodes. In particular, a second-order structure is

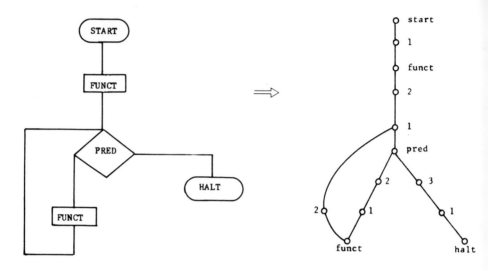

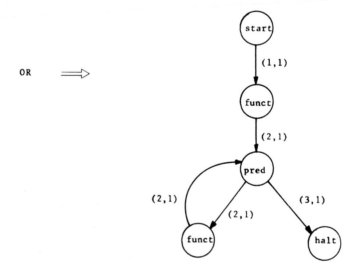

Fig. 3.11. Transformation from plex structure to graphs.

called a branch structure and a first-order structure a node structure. Then an mth-order context-free graph grammar G_g is a quadruple $G_g = (V_N, V_T, P, S)$ where V_N is a set of mth-order nonterminal structures: nodes, branches, triangles, ..., polygons with m vertices; V_T is a set of terminals: nodes and branches; P is a finite set of productions of the form $A \to \alpha$, where A is a nonterminal structure and α a graph containing possibly both terminals and nonterminals (α is connected to the rest of the graph through exactly the same nodes as A); S is a set of initial graphs.

The expression $A * B$ denotes that the two graphs A and B are connected by a pair of nodes (Fig. 3.12a), and $N(A + B + C)$ denotes that the graphs

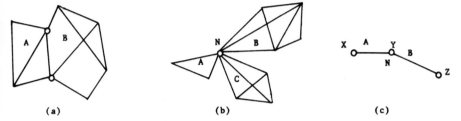

(a)   (b)   (c)

Fig. 3.12.   Illustrations of (a) $A * B$, (b) $N(A + B + C)$, and (c) ANB.

A, B, and C are connected through a common node N (Fig. 3.12b).† Thus the production $A \to B * C$, where A, B, and C are branch structures, should be interpreted as: Replace branch structure A by branch structures B and C connected to the graph through the same nodes as A. No other connection exists between B and C. Similarly, the production $N \to M(A + B)$ should be interpreted as: Replace node structure N by a node structure M and two other structures A and B connected to the rest of the graph by the same node as N. When no ambiguity occurs, we can use simple concatenations, for example, ANB to denote a nonterminal subgraph consisting of a branch structure A with nodes X and Y connected to the node structure N through Y and a branch structure B with nodes Y and Z connected to N through Y (Fig. 3.12c). The subgraph is connected to the rest of the graph through the nodes X and Z. The following examples illustrate the use of graph grammars for pattern description.

Example 3.8   The following grammar describes graphs representing two-terminal series-parallel networks (TTSPN):

$$G_g = (V_N, V_T, P, S)$$

---

† The expressions $A * B$ and $N(A + B + C)$ are defined as shorthand notations and they are not necessarily part of the vocabulary.

where

$$V_N = \{S, B\}, \qquad V_T = \{\overset{b}{-}, \overset{n}{\cdot}\}, \qquad S = \{nBn\}$$

and P:

(1) $B \to BnB$

(2) $B \to B * B$

(3) $B \to b$

A typical generation would be

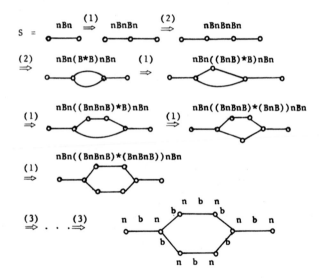

Example 3.9 The following graph grammar describes connected trees:

$$G_g = (V_N, V_T, P, S)$$

where

$$V_N = \{N\}, \qquad V_T = \{\overset{b}{-} \overset{n}{\cdot}\}, \qquad S = \{N\}$$

and P:

$$N \to NbN$$

$$N \to n$$

A typical generation would be

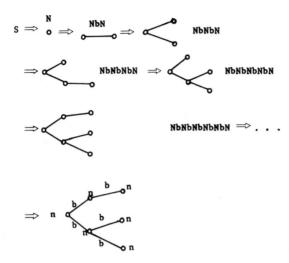

Example 3.10   The set of all convex polygons can be generated by the graph grammar $G_g = (V_N, V_T, P, S)$ where $V_N = \{S, E\}$, $V_T = \{e\}$, and P:

For generating all polygons, replace (2) by

By the extension of one-dimensional concatenation to multidimensional contatenation, strings are generalized to trees. Tree grammars and the corresponding recognizers, tree automata, have been studied recently by a number of authors [75, 76]. Naturally, if a pattern can be conveniently described by a tree, it will easily be generated by a tree grammar. For example, in Fig. 3.13, patterns and their corresponding tree representations are listed in parts (a) and (b), respectively. More detailed discussions about tree grammars, tree automata, and their applications to syntactic pattern recognition are included in Appendix I.

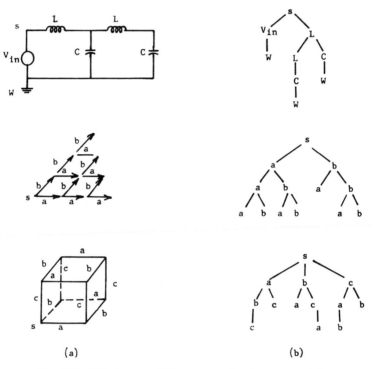

(a)   (b)

Fig. 3.13. (a) Patterns and (b) corresponding tree representations.

3.4 The Use of Semantic Information

In specifying a selected primitive, a set of attributes is usually required. A primitive with different properties can be expressed in terms of its attribute values. This set of attributes may be considered as the semantic information of the primitive. Each attribute may be expressed by numerical values or logical predicates [49, 54]. For example, the coordinates of head and tail can be used to specify the exact location of a line segment, or the range of its radius can be used to specify a primitive "arc." Anderson has proposed the use of six spatial coordinates in describing each character (primitive)† for the recognition of two-dimensional mathematical expressions [64]. Following Shaw's formalism [65], a primitive description of a pattern x, $T(x)$, is defined as

$$T(x) = (T_s(x), T_v(x)) \tag{3.11}$$

where $T_s(x)$ specifies the primitives contained in the pattern x and their relationship to one another, and $T_v(x)$ gives the attribute values of each primitive in x. With respect to the structural (syntactic) information, x is described by a syntactic or hierarchic description

$$H(x) = (H_s(x), H_v(x)) \tag{3.12}$$

where $H_s(x)$, representing the syntactic component, consists of the set of productions of the grammar G that may be used to generate $T_s(x)$, and $H_v(x)$ is the result of obeying the corresponding interpretation rule for each production of G used in parsing $T_s(\alpha)$. Thus, a complete description of a pattern x is defined as

$$(T(x), H(x)) = ((T_s(x), T_v(x)), (H_s(x), H_v(x))) \tag{3.13}$$

A class of patterns can be described by a grammar G such that for each pattern x in the class $T_s(x) \in L(G)$, and by the associated semantic information. It should be noted that the "meaning" of a pattern is expressed by both its primitive and its syntactic descriptions. Several grammars may be used to generate the same class of primitive descriptions, but the syntactic descriptions may be different for different grammars. More generally, the same class of patterns may be described by totally different primitive and syntactic descriptions; the intended interpretation of the picture dictates its description.

The semantic information of a subpattern (nonterminal) is, in general, evaluated either from the semantic information of the composed primitives according to the syntactic relations or operators and the semantic rules associated with each production of the grammar, or on the basis of a separate set of functions or rules which are not necessarily defined in conjunction with the productions of the grammar.‡

---

† Similar approaches are also proposed by Chang [66], Nake [67], and Simon and Checroun [68, 69]. This class of grammars is sometimes called "grammars with coordinates" [70].

‡ Refer to Section 1.4.1 for some preliminary discussions on the use of picture properties to characterize further each subpicture or primitive. Recently, based on the theory of systemic grammar, Winograd has developed a system for language understanding, in which considerable interplay between syntax and semantics is permitted. The application of the language-understanding system to scene analysis has been demonstrated [71].

Example 3.11   The meanings of the binary concatenation operators in PDL are presented in terms of the semantic function. CAT means "is concatenated onto."

$$T_v(A) = (S_1(tl(a), hd(a))\ S_2(tl(b), hd(b))) \qquad (3.14)$$

$T_s(A)$	Production	Semantic Function
$S_1 + S_2$	$A \to S_1 + S_2$	$tl(A) = tl(S_1)$
		$hd(A) = hd(S_2)$
		and $hd(S_1)$ CAT $tl(S_2)$
$S_1 \times S_2$	$A \to S_1 \times S_2$	$tl(A) = tl(S_1)$
		$hd(A) = hd(S_2)$
		and $tl(S_1)$ CAT $tl(S_2)$
$S_1 - S_2$	$A \to S_1 - S_2$	$tl(A) = tl(S_1)$
		$hd(A) = hd(S_2)$
		and $hd(S_1)$ CAT $hd(S_2)$
$S_1 * S_2$	$A \to S_1 * S_2$	$tl(A) = tl(S_1)$
		$hd(A) = hd(S_2)$
		and $(tl(S_1)$ CAT $tl(S_2))$
		$\wedge$ $(hd(S_1)$ CAT $hd(S_2))$

Example 3.12   Referring to Fig. B-3 in Appendix B, for the patterns "A" and "HOUSE," we see that

$T_s(A) = (b + (((b + c) * a) + c))$
$H_s(A)$ is represented by the tree (parse) describing "A" in Fig. B-3;
$H_v(A) = ((S, (tl, hd)_S), (A, (tl, hd)_A), (TRIANGLE, (tl, hd)_{TRIANGLE}),$
$\qquad\qquad (ANGLE, (tl, hd)_{ANGLE}))$

where $(tl, hd)_N$ is the tail and head of the nonterminal N.

$T_s(HOUSE) = ((e + (a + (\sim e))) * ((b + c) * a))$
$H_v(HOUSE)$ is represented by the tree (parse) describing "House" in
$\qquad\qquad$ Fig. B-3.

$H_v(HOUSE)\ = ((S, (tl, hd)_S), (HOUSE, (tl, hd)_{HOUSE}),$
$\qquad\qquad\qquad (TRIANGLE, (tl, hd)_{TRIANGLE}), (ANGLE, (tl, hd)_{ANGLE}))$

Knuth has recently formalized the semantics of context-free languages [72]. His approach is to define the "meaning" of a string in terms of attributes for each of the nonterminals that arise when the string is parsed according to the grammatical rules. The attributes of each nonterminal correspond to the meaning of the phrase produced from that nonterminal. Attributes are classified into two kinds, "inherited" and "synthesized." Inherited attributes are those aspects of meaning which come from the context of a phrase, whereas synthesized ones are those aspects which are built up from within the phrase. There can be considerable interplay between inherited and synthesized attributes; the essential idea is that the meaning of an entire string is built up from local rules relating the attributes of each production appearing in the parse of that string. For each production in a context-free grammar, "semantic rules" are specified which define (i) all of the synthesized attributes of the nonterminal on the left side of the production, and (ii) all of the inherited attributes of the nonterminals on the right side of the production. The initial nonterminal S (at the root of the parse tree) has no inherited attributes.

In a context-free language, in principle, synthesized attributes alone are sufficient to define the meaning of a string, but inclusion of inherited attributes often leads to important simplifications in the attribute evaluation. Naturally, if the grammar is ambiguous, the meanings obtained by applying the semantic rules to the different derivation trees of a string may not be the same.

Semantic rules are added to a context-free grammar $G = (V_N, V_T, P, S)$ in the following way. To each symbol $X \in (V_N \cup V_T)$, associate a finite set $A(X)$ of attributes; and $A(X)$ is partitioned into two disjoint sets, the synthesized attribute set $A_0(X)$ and the inherited attribute set $A_1(X)$. We require that $A_1(S) = \varnothing$ and, similarly, that $A_0(x) = \varnothing$ if $X \in V_T$. Each attribute $\alpha \in A(X)$ has a set of possible values D_α, from which one value will be selected for each appearance of X in a derivation tree. Let P consist of m productions, and let the kth production be

$$X_{k0} \to X_{k1} X_{k2} \cdots X_{kn_k} \tag{3.15}$$

where $X_{k0} \in V_N$, $n_k \geqslant 0$, and $X_{kj} \in (V_N \cup V_T)$, $1 \leqslant j \leqslant n_k$. The semantic rules are functions $f_{kj\alpha}$ defined for all $1 \leqslant k \leqslant m$, $0 \leqslant j \leqslant n_k$, and $\alpha \in A_0(X_{kj})$ if $j = 0$, $\alpha \in A_1(X_{kj})$ if $j > 0$. Each such function is a mapping of $D_{\alpha_1} \times D_{\alpha_2} \times \cdots \times D_{\alpha_t}$ into D_α for some $t = t(k, j, \alpha) \geqslant 0$, where each $\alpha_i = \alpha_i(k, j, \alpha)$, $1 \leqslant i \leqslant t$, is an attribute of some X_{kl_i} for $0 \leqslant l_i \leqslant n_k$. In other words, each semantic rule maps values of certain attributes of $X_{k0}, \ldots, X_{kn_k}$ into the value of some attribute of X_{kj}.

The semantic rules are used to assign a "meaning" to strings of a context-free language in the following way. For any derivation of a terminal string x from S by a sequence of productions, construct the derivation tree in the usual way. The root of this tree is S and each node is labeled either with a terminal symbol, or with a nonterminal symbol X_{k0} corresponding to an application of the kth production, for some k. Let X be the label of a node of the tree and let $\alpha \in A(X)$ be an attribute of X. If $\alpha \in A_0(X)$, then $X = X_{k0}$ for some k; while if $\alpha \in A_1(X)$, then $X = X_{kj}$ for some j and k, $1 \leqslant j \leqslant n_k$. The attribute α is defined to have the value v at this node if, in the corresponding semantic rule

$$f_{kj\alpha}: \quad D_{\alpha_1} \times \cdots \times D_{\alpha_t} \to D_\alpha \tag{3.16}$$

all of the attributes have the respective values $v_1, \ldots, v_t$ at the respective nodes labeled $X_{kl_1}, \ldots, X_{kl_t}$, and $v = f_{kj\alpha}(v_1, \ldots, v_t)$. This process of attribute evaluation is to be applied throughout the tree until no more attribute values can be defined, at which point the evaluated attributes at the root of the tree constitute the "meaning" corresponding to the derivation tree.

Example 3.13   The meaning of binary notation for a number can be evaluated at the following two different ways. Given a grammar $G = (V_N, V_T, P, S)$, where $V_N = \{S, B, L\}$, $V_T = \{0, 1, \cdot\}$, and P:

$$\begin{array}{ll} S \to L, & S \to L \cdot L \\ L \to B, & L \to LB \\ B \to 0, & B \to 1 \end{array}$$

The derivation tree of the string 1101.01 is

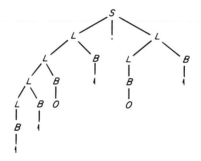

(1)   With synthesized and inherited attributes:
The attributes are $A_0(B) = \{v\}$, $A_1(B) = \{s\}$, $A_0(L) = \{v, l\}$, $A_1(L) = \{s\}$, $A_0(S) = \{v\}$, $A_1(S) = \varnothing$, and $A_0(a) = A_1(a) = \varnothing$ for $a \in V_T$. The attribute value sets are $D_v = \{\text{rational numbers}\}$, $D_s = D_l = \{\text{integers}\}$.

Each B has a "value" $v(B)$ which is a rational number.
Each B has a "scale" $s(B)$ which is an integer.
Each L has a "value" $v(L)$ which is a rational number.
Each L has a "length" $l(L)$ which is an integer.
Each L has a "scale" $s(L)$ which is an integer.
Each S has a "value" $v(S)$ which is a rational number.

Production	Semantic Rule
$B \rightarrow 0$	$v(B) = 0$
$B \rightarrow 1$	$v(B) = 2^{s(B)}$
$L \rightarrow B$	$v(L) = v(B), \quad s(B) = s(L), \quad l(L) = 1$
$L_1 \rightarrow L_2\, B$	$v(L_1) = v(L_2) + v(B), \quad s(B) = s(L_1)$
	$s(L_2) = s(L_1) + 1, \quad l(L_1) = l(L_2) + 1$
$S \rightarrow L$	$v(S) = v(L), \quad s(L) = 0$
$S \rightarrow L_1 \cdot L_2$	$v(S) = v(L_1) + v(L_2), \quad s(L_1) = 0$
	$s(L_2) = -l(L_2)$

Thus, the derivation tree of 1101.01 with the semantic information is

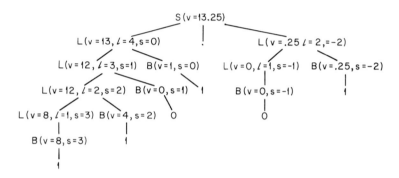

(2)   With synthesized attributes:

Each B has a "value" $v(B)$ which is an integer.
Each list of bits L has a "length" $l(L)$ which is an integer.
Each list of bits L has a "value" $v(L)$ which is an integer.
Each number S has a "value" $v(S)$ which is a rational number.

Production	Semantic Rule
$B \rightarrow 0$	$v(B) = 0$
$B \rightarrow 1$	$v(B) = 1$
$L \rightarrow B$	$v(L) = v(B), \quad l(L) = 1$
$L_1 \rightarrow L_2\, B$	$v(L_1) = 2^{v(L_2)} + v(B), \quad l(L_1) = l(L_2) + 1$
$S \rightarrow L$	$v(S) = v(L)$
$S \rightarrow L_1 \cdot L_2$	$v(S) = v(L_1) + v(L_2)/2^{l(L_2)}$

Thus, the derivation tree of 1101.01 with the semantic information is

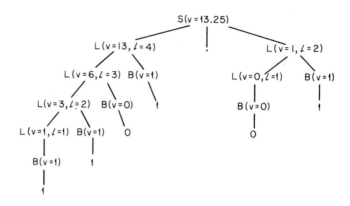

Example 3.14†   This example illustrates the application of Knuth's formal semantics to patterns described by PDL expressions. Let $C(X)$ denote the connectivity of X. The attributes of X in this case are $\text{tl}(X)$, $\text{hd}(X)$, and $C(X)$. λ is the " null " primitive. Referring to Appendix B, for English characters E and N, the evaluation of semantic information may be carried out according to the following systematic procedure.

(1)   Character E

Production	*Semantic Rule*
$E1 \rightarrow (v2 + h2)$	$\text{tl}(E1) = \text{tl}(v2)$
	$\text{hd}(E1) = \text{hd}(h2)$
	$C(E1) = \text{hd}(v2) \text{ CAT } \text{tl}(h2)$
$E2 \rightarrow (E1 \times h1)$	$\text{tl}(E2) = \text{tl}(E1)$
	$\text{hd}(E2) = \text{hd}(E1)$
	$C(E2) = C(E1) \wedge (\text{tl}(E1) \text{ CAT } \text{tl}(h1))$
$E3 \rightarrow (v2 + E2)$	$\text{tl}(E3) = \text{tl}(v2)$
	$\text{hd}(E3) = \text{hd}(E2)$
	$C(E3) = (\text{hd}(v2) \text{ CAT } \text{tl}(E2)) \wedge C(E2)$
$E \rightarrow (E3 \times h2)$	$\text{tl}(E) = \text{tl}(E3)$
	$\text{hd}(E) = \text{hd}(h2)$
	$C(E) = C(E3) \wedge (\text{tl}(E3) \text{ CAT } \text{tl}(h2))$

† This example was provided by Mr. K. Tachibana of the University of California at Berkeley.

The derivation tree of E with the semantic information is

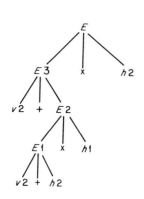

$$\text{tl}(E) = \text{tl}(v2), \quad \text{hd}(E) = \text{hd}(h2)$$
$$C(E) = ((\text{hd}(v2) \text{ CAT } \text{tl}(v2)) \wedge$$
$$((\text{hd}(v2) \text{ CAT } \text{tl}(h2)) \wedge$$
$$(\text{tl}(v2) \text{ CAT } \text{hd}(h1)))) \wedge$$
$$(\text{tl}(v2) \text{ CAT } \text{tl}(h2))$$
$$\text{tl}(E3) = \text{tl}(v2), \quad \text{hd}(E3) = \text{hd}(h1)$$
$$C(E3) = (\text{hd}(v2) \text{ CAT } \text{tl}(v2)) \wedge C(E2)$$
$$\text{tl}(E2) = \text{tl}(v2), \quad \text{hd}(E2) = \text{hd}(E1)$$
$$C(E2) = C(E1) \wedge (\text{tl}(v2) \text{ CAT } \text{tl}(h1))$$
$$\text{tl}(E1) = \text{tl}(v2), \quad \text{hd}(E1) = \text{hd}(h2)$$
$$C(E1) = \text{hd}(v2) \text{ CAT } \text{tl}(h2)$$
$$\text{tl}(E1) = \text{tl}(v2), \quad \text{hd}(E1) = \text{hd}(h2)$$
$$C(E1) = \text{hd}(v2) \text{ CAT } \text{tl}(h2)$$

(2)   Character N

Production	Semantic Rule
$N1 \rightarrow (v3 \times \lambda)$	$\text{tl}(N1) = \text{tl}(v3)$
	$\text{hd}(N1) = \text{hd}(\lambda) = \text{tl}(\lambda)$
	$C(N1) = \text{tl}(v3) \text{ CAT } \text{tl}(\lambda)$
$N2 \rightarrow (v3 + g3)$	$\text{tl}(N2) = \text{tl}(v3)$
	$\text{hd}(N2) = \text{hd}(g3)$
	$C(N2) = \text{hd}(v3) \text{ CAT } \text{tl}(g3)$
$N \rightarrow (N2 + N1)$	$\text{tl}(N) \ = \text{tl}(N2)$
	$\text{hd}(N) = \text{hd}(N1)$
	$C(N) = C(N2) \wedge (\text{hd}(N2) \text{ CAT }$
	$\text{tl}(N1)) \wedge C(N1)$

The derivation tree of N with the semantic information is

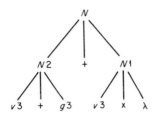

$$tl(N) = tl(v3), \quad hd(N) = hd(\lambda)$$
$$C(N) = (hd(v3) \text{ CAT } tl(g3)) \wedge$$
$$(hd(g3) \text{ CAT } tl(v3)) \wedge$$
$$(tl(v3) \text{ CAT } tl(\lambda))$$
$$tl(N1) = tl(v3), \quad hd(N1) = hd(\lambda)$$
$$C(N1) = tl(v3) \text{ CAT } tl(\lambda)$$
$$tl(N2) = tl(v3), \quad hd(N2) = hd(g3)$$
$$C(N2) = hd(v3) \text{ CAT } tl(g3)$$

References

1. W. Stallings, Recognition of printed Chinese characters by automatic pattern analysis. *Comp. Graphics and Image Process.* **1**, 47–65 (1972).
2. T. Sakai, M. Nagas and H. Terai, A description of Chinese characters using subpatterns. *Information Processing in Japan* **10**, 10–14, (1970).
3. S. K. Chang, An interactive system for Chinese characters generation and text editing. *Proc. Int. Conf. Cybernet. and Soc. Washington, D. C., October 9–12, 1972*, pp. 206–214.
4. R. L. Grimsdale, F. H. Summer, C. J. Tunis, and T. Kilburn, A system for the automatic recognition of patterns. *Proc. IEE* **106**, Pt. B, 210–221 (1959). Reprinted in *Pattern Recognition* (L. Uhr, ed.), pp. 317–318. Wiley, New York, 1966.
5. M. Eden and M. Halle, The characterization of cursive writing. *Proc. Symp. Inform. Theory, 4th, London* (C. Cherry, ed.), pp. 287–299. Butterworth, London, 1961.
6. L. D. Earnest, Machine recognition of cursive writing. *Inform. Process. Proc. IFIP (Int. Fed. Inform. Process.) Congr. 1962*, pp. 462–466. North-Holland Publ. Amsterdam, 1963.
7. P. Mermelstein and M. Eden, Experiments on computer recognition of connected handwritten words. *Information and Control* **7**, 255–270 (1964).
8. M. Eden and P. Mermelstein, Models for the dynamics of handwriting generation. *Proc. Annu. Conf. Eng. Med. and Biol., 16th, Baltimore, Maryland, 1963*, pp. 12–13.
9. H. Freeman, On the encoding of arbitrary geometric configurations. *IEEE Trans. Electron. Comput.* **EC-10**, 260–268 (1961).
10. H. Freeman On the digital-computer classification of geometric line patterns. *Proc. Nat. Electron. Conf.* **18**, 312–324 (1962).
11. P. J. Knoke and R. G. Wiley, A linguistic approach to mechanical pattern recognition. *Proc. IEEE Comput. Conf. September 1967*, pp. 142–144.
12. J. Feder, Languages of encoded line patterns. *Information and Control* **13**, 230–244 (1968).
13. N. Chomsky, A transformational approach to syntax. In *The Structure of Language* (J. A. Fodor and J. J. Katz, eds.). Prentice-Hall, Englewood Cliffs, New Jersey, 1964.
14. E. Bach, *An Introduction to Transformation Grammars*. Holt, New York, 1964.

15. G. Gallus and P. Nurath, Improved computer chromosome analysis incorporating preprocessing and boundary analysis. *Phys. Med. Biol.* **15**, 435–445 (1970).
16. H. C. Lee and K. S. Fu, *Stochastic Linguistics for Picture Recognition.* Tech. Rep. TR-EE 72-17. School of Elec. Eng., Purdue Univ., Lafayette, Indiana, 1972.
17. H. Freeman and S. P. Morse, On searching a contour map for a given terrain elevation profile. *J. Franklin Inst.* **284**, 1–25 (1967).
18. J. Feder and H. Freeman, Digital curve matching using a contour correlation algorithm. *IEEE Int. Conv. Rec.* Pt. 3, 69–85 (1966).
19. C. T. Zahn, *Two-Dimensional Pattern Description and Recognition via Curvature Points.* SLAC Rep. No. 72. Stanford Linear Accelerator Center, Stanford, California, 1966.
20. H. Freeman and J. Garder, Pictorial jig-saw puzzles: The computer solution of a problem in pattern recognition. *IEEE Trans. Electron. Comput.* **EC-13**, 118–127 (1964).
21. R. Narasimhan, Syntax-directed interpretation of classes of pictures. *Comm. ACM* **9**, 166–173 (1966).
22. R. J. Spinrad, Machine recognition of hand printing. *Information and Control* **8**, 124–142 (1965).
23. J. F. O'Callaghan, Problems in on-line character recognition. In *Picture Language Machines* (S. Kaneff, ed.). Academic Press, New York, 1970.
24. M. Nir, *Recognition of General Line Patterns with Application to Bubble-Chamber Photographs and Handprinted Characters.* Ph.D. Thesis, Moore School of Elec. Eng. Univ. of Pennsylvania, Philadelphia, Pennsylvania, 1967.
25. A. C. Shaw, *The Formal Description and Parsing of Pictures.* SLAC Rep. No. 84. Stanford Linear Accelerator Center, Stanford, California, 1968.
26. R. Narasimhan, Labeling schemata and syntactic description of pictures. *Information and Control* **7**, 151–179 (1964).
27. J. W. Butler, M. K. Butler, and A. Stroud, Automatic classification of Chromosomes. *Proc. Conf. Data Acquisition and Process. Biol. Med. New York, 1963.*
28. R. S. Ledley *et al.*, FIDAC: Film input to digital automatic computer and associated syntax-directed pattern-recognition programming system. In *Optical and Electro-Optical Information Processing* (J. T. Tippet, D. Beckowitz, L. Clapp, C. Koester, and A. Vanderburgh, Jr., eds.). pp. 591–613 MIT Press, Cambridge, Massachusetts, 1965.
29. W. J. Hankley and J. T. Tou, Automatic fingerprint interpretation and classification via contextual analysis and topological Coding. In *Pictorial Pattern Recognition* (G. C. Cheng, R. S. Ledley, D. K. Pollock, and A. Rosenfeld, eds.). Thompson Book Co., Washington, D.C., 1968.
30. A. Grasselli, On the automatic classification of fingerprints. In *Methodologies of Pattern Recognition* (S. Watanabe, ed.). Academic Press, 1969. New York.
31. G. Levi and F. Sirovich, Structural description of fingerprint images. *Information Sci.* **4**, 327–356 (1972).
32. B. Moayer and K. S. Fu, *A Syntactic Approach to Fingerprint Pattern Recognition.* Tech. Rep. TR-EE 73-18. School of Elec. Eng., Purdue Univ., West Lafayette, Indiana, 1973.
33. M. Nagao, Picture recognition and data structure. In *Graphic Languages* (F. Nake and A. Rosenfeld, eds.). North-Holland Publ. Amsterdam, 1972.
34. M. D. Kelley, *Visual Identification of People by Computer.* Ph.D. Thesis, Dept. of Comput. Sci., Stanford Univ., Stanford, California, 1970.
35. L. G. Roberts, Machine perception of three-dimensional solids. In *Optical and Electro-Optical Information Processing*, (J. T. Tippet, D. Beckowitz, L. Clapp, C. Koester, and A. Vanderburgh, Jr., eds.). pp. 159–197. MIT Press, Cambridge, Massachusetts, 1965.

36. R. O. Duda and P. E. Hart, Experiments in scene analysis. *Proc. Nat. Symp. Ind. Robots, 1st, Chicago, April 1970.*
37. J. A. Feldman *et al.*, The Stanford hand-eye project. *Proc. Int. Joint Conf. Artificial Intelligence, 1st, Washington, D. C., May 1969.*
38. T. Pavlidis, Analysis of set patterns. *Pattern Recognition* 1, 165–178 (1968).
39. J. L. Kelley, *General Topology.* Van Nostrand-Reinhold, Princeton, New Jersey, 1955.
40. A. Rosenfeld and J. P. Strong, A grammar for maps. In *Software Engineering* (J. T. Tou, ed.), Vol. 2. Academic Press, New York, 1971
41. T. Pavlidis, Representation of figures by labeled graphs. *Pattern Recognition* 4, 5–17 (1972).
42. T. Pavlidis, Structural pattern recognition; Primitives and juxtaposition. In *Frontiers of Pattern Recognition* (S. Watanabe, ed.). Academic Press, New York, 1972.
43. R. O. Duda and P. E. Hart, *Pattern Classification and Scene Analysis.* Artificial Intelligence Group, Stanford Res. Inst., Menlo Park, California, 1970.
44. M. L. Minsky and S. Papert, Proj. MAC Progr. Rep. IV. MIT Press, Cambridge, Massachusetts, 1967.
45. A. Guzman, Decomposition of scenes into bodies. *Proc. AFIPS Fall Joint Comput. Conf., San Francisco, California, 1968,* 33, Pt. 1, pp. 291–304.
46. C. R. Brice and C. L. Fennema, Scene analysis using regions. *Artificial Intelligence* 1, 205–226 (1970).
47. M. C. Clowes, Transformational grammars and the organization of pictures. In *Automatic Interpretation and Classification of Images* (A. Grasselli, ed.). Academic Press, New York, 1969.
48. L. Kanal and B. Chandrasekaran, On linguistic, statistical and mixed models for pattern recognition. In *Frontiers of Pattern Recognition* (S. Watanabe, ed.). Academic Press, New York, 1972.
49. R. Narasimhan, On the description, generation, and recognition of classes of pictures. In *Automatic Interpretation and Classification of Images* (A. Grasselli, ed.). Academic Press, New York, 1969.
50. R. Narasimhan, Picture languages. In *Picture Language Machines* (S. Kaneff, ed.). Academic Press. New York, 1970.
51. T. G. Evans, A formalism for the description of complex objects and its implementation. *Proc. Int. Congr. Cybernet. 5th, Namur, Belgium, September 1967.*
52. M. L. Minsky, Steps toward artificial intelligence. *Proc. IRE* 49, 8–30 (1961).
53. M. B. Clowes, Pictorial relationships—A syntactic approach. In *Machine Intelligence 4,* (B. Meltzer and D. Michie, eds.). Amer. Elsevier, New York, 1969.
54. T. G. Evans, Descriptive pattern analysis techniques. In *Automatic Interpretation and Classification of Images* (A. Grasselli, ed.). Academic Press, New York, 1969.
55. H. G. Barrow and J. R. Popplestone, Relational descriptions in picture processing. In *Machine Intelligence 6* (B. Meltzer and D. Michie, eds.). pp. 377–396. Edinburgh Univ. Press, Edinburgh, 1971.
55a. R. S. Ledley, *Programming and Utilizing Digital Computers.* McGraw-Hill, New York, 1962.
56. R. A. Kirsch, Computer interpretation of English text and patterns. *IEEE Trans. Electron. Comput.* EC-13, 363–376 (1964).
57. M. F. Dacey, The syntax of a triangle and some other figures. *Pattern Recognition* 2, 11–31 (1970).
58. D. M. Milgram and A. Rosenfeld, Array automata and array grammars. *IFIP Congr. 71,* Booklet TA-2, pp. 166–173. North-Holland Publ. Amsterdam, 1971.

59. J. L. Pfaltz and A. Rosenfeld, Web grammars. *Proc. Int. Joint Conf. Artificial Intelligence, 1st, May 1969, Washington, D. C.* pp. 609–619.

60. A. C. Shaw, Picture graphs, grammars, and parsing. In *Frontiers of Pattern Recognition* (S. Watanabe, ed.). Academic Press, New York, 1972.

61. A. Rosenfeld, Picture automata and grammars: An annotated bibliography. *Proc. Symp. Comput. Image Process. Recognition, Univ. of Missouri, Columbia, August 24–26, 1972,* Vol. 2.

62. T. Pavlidis, Linear and context-free graph grammars. *J. Assoc. Comput. Mach.* **19**, 11–22 (1972).

63. T. Pavlidis, Graph theoretic analysis of pictures. In *Graphic Languages* (F. Nake and A. Rosenfeld, eds.). North-Holland Publ. Amsterdam, 1972.

64. R. H. Anderson, *Syntax-Directed Recognition of Hand-Printed Two-Dimensional Mathematics.* Ph.D. Thesis, Harvard Univ., Cambridge, Massachusetts, 1968.

65. A. C. Shaw, The formal picture description scheme as a basis for picture processing systems. *Information and Control* **14**, 9–52 (1969).

66. S. K. Chang, Picture processing grammar and its applications. *Information Sci.* **3**, 121–148 (1971).

67. F. Nake, A proposed language for the definition of arbitrary two-dimensional signs. In *Pattern Recognition in Biological and Technical Systems* (O. J. Grusser and R. Klinke, eds.). Springer-Verlag, Berlin and New York, 1971.

68. J. C. Simon and A. Checroun, Pattern linguistic analysis invariant for plane transformations. *Proc. Int. Joint. Conf. Artificial Intelligence, 2nd, London, September 1971,* pp. 308–317.

69. J. C. Simon, A. Checroun, and C. Roche, A method of comparing two patterns independent of possible transformations and small distortions. *Pattern Recognition* **4**, 73–82 (1972).

70. D .L. Milgram and A. Rosenfeld, A note on "Grammars with Coordinates". In *Graphic Languages* (F. Nake and A. Rosenfeld, eds.). North-Holland Publ., Amsterdam, 1972.

71. M. L. Minsky and S. Papert, *Research at the Laboratory in Vision, Language, and Other Problems of Intelligence.* AI Memo. No. 252, Proj. MAC. MIT. Press, Cambridge, Massachusetts, 1972.

72. D. E. Knuth, Semantics of context-free languages. *J. Math. Syst. Theory* **2**, 127–146 (1968).

73. K. S. Fu and P. H. Swain, On syntactic pattern recognition. In *Software Engineering* (J. T. Tou, ed.), Academic Press, New York, 1971.

74. S. Ginsburg and B. Partee, A mathematical model of transformation grammars. *Information and Control* **15**, 297–234 (1969).

75. W. S. Brainerd, Tree generating regular systems. *Information and Control* **14**, 217–231 (1969).

76. J. E. Donar, Tree acceptors and some of their applications. *J. Comput. System Sci.* **4**, 406–451 (1970).

77. J. J. Katz and J. A. Fodor, The structure of a semantic theory. *Language* **39**, 170–210 (1963).

78. A. A. Letichevskii, Syntax and semantics of formal languages. *Kibernetika (Kiev)* **4**, 1–9 (1968).

79. S. Kaneff, *Picture Language Machines.* Academic Press, New York, 1970.

80. J. Myloponlos, On the application of formal language and automata theory to pattern recognition. *Pattern Recognition* **4**, 37–52 (1972).

81. K. J. Kreeding and J. O. Amoss, A pattern description language-PADEL. *Pattern Recognition* **4**, 19–36 (1972).

82. M. B. Clowes, Scene analysis and picture grammars. In *Graphic Languages* (F. Nake and A. Rosenfeld, eds.). North-Holland Publ., Amsterdam, 1972.

83. T. Pavlidis, Computer recognition of figures through decomposition. *Information and Control* **12**, 526–537 (1968).

84. I. B. Muchnik, Simulation of process of forming the language for description and analysis of the forms of images. *Pattern Recognition* 101–140 (1972).

85. A. Rosenfeld, Isotonic grammars, parallel grammars, and picture grammars. In *Machine Intelligence 6* (B. Meltzer and D. Michie, eds.), pp. 281–294. Edinburgh Univ. Press, Edinburgh, 1971.

86. D. F. Londe and R. F. Simmons, NAMER: A pattern recognition system for generating sentences about relations between line drawings. *Proc. ACM Nat. Conf., 20th, 1965*, pp. 162–175.

87. A. Inselbey and R. Kline, *SAP: A Model for the Syntactic Analysis of Pictures.* Tech. Rep. No. 9. Comput. Syst. Lab., Washington Univ., June 1968.

88. P. D. Vigna and F. Luccio, Some aspects of the recognition of convex polyhedra from two plane projections. *Information Sci.* **2**, 159–178 (1970).

89. D. A. Huffman, Impossible objects as nonsense sentences. In *Machine Intelligence 6* (B. Meltzer and D. Michie, eds.), pp. 113–125. Edinburgh Univ. Press, Edinburgh, 1971.

90. R. Narasimhan and V. S. N. Reddy, A syntax-aided recognition scheme for hand-printed English letters. *Pattern Recognition* **3**, 344–362 (1971).

91. F. P. Preparata and S. R. Ray, An approach to artificial nonsymbolic cognition. *Information Sci.* **4**, 65–86 (1972).

Chapter 4

Syntax Analysis as a Recognition Procedure

4.1 Introduction

After a grammar is constructed to generate a language which would describe the pattern under study, the next step is to design a recognizer that will
recognize the patterns (represented by strings) generated by the grammar. A
straightforward approach might be to construct a grammar for each class of
patterns.† The recognizer designed for a particular grammar will recognize
only the patterns in the class corresponding to this grammar. For example,
corresponding to m classes of patterns $\omega_1, \omega_2, \ldots, \omega_m$, we can construct m
grammars $G_1, G_2, \ldots, G_m$, respectively, such that the strings generated by
the grammar G_i would represent the patterns in class ω_i. For an unknown
pattern described by the string x, the problem of recognition is essentially reduced to the answer to the question:

$$\text{Is} \quad x \in L(G_i) \quad \text{for} \quad i = 1, \ldots, m?$$

The process that would result in an answer to such a question with respect to
a given grammar is, in general, called "syntax analysis" or "parsing" [1–8].
In addition to giving an answer of "yes" or "no" to such a question, the

---

† As can be seen in later chapters, the approach of constructing one grammar for each
pattern class may not always be necessary.

91

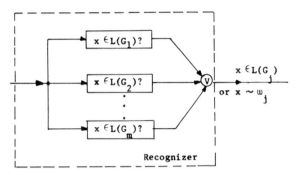

Fig. 4.1. Block diagram of the syntactic recognizer.

process can also provide the generation or derivation tree of x, the structural information of the corresponding pattern. A block diagram for the recognizer is shown in Fig. 4.1. Presumably, if different patterns are uniquely represented by different strings, only one answer to the m questions "Is $x \in L(G_i)$, $i = 1, \ldots, m$?" will be yes, the rest will be no. If the answer is $x \in L(G_j)$, then x is recognized from the class ω_j. If all the answers are no, the pattern x will be rejected; that is, x is not accepted as from any one of the m classes.

A very simple form of implementing the recognizer is to match the input string x with some prototype or reference strings from each class in terms of pattern primitives [9–11]. The matching of primitives can be performed in parallel or sequentially. x will be recognized as from the same class of the reference string that " best " matches x. In this case, either a perfect match or an appropriately selected matching criterion is required. The choice of reference strings is of course also very important. This approach, though it seems less flexible, is simple and speedy in processing. However, the syntactic (hierarchical structural) information of the string has really not been sufficiently utilized in the recognition, and the approach is useful only when appropriate reference strings and a meaningful matching criterion can be determined.†

If the grammar G_i is finite state, a deterministic finite-state automaton can be constructed to recognize the strings generated by G_i [see Chapter 2]. If G_i is a context-free or context-free programmed grammar, a nondeterministic automaton is usually required. The process or the algorithm of performing the recognition is called " syntax analysis," which uses, in general, a non-

---

† An obvious extension of this approach to include syntactic information is to match the graph (e.g., relational graph) describing the unknown pattern (or subgraphs) against various prototype or reference graphs, or to match the derivation tree (parse) of the unknown string x (or subtrees) against reference parses (e.g., see Appendix E) [12–19]. It is also possible, for some problems, to set up a group of tests which determine whether or not a pattern has certain attributes and/or what the actual values of the attributes are. Based on the outcome of these tests, the pattern can be assigned to one of the predetermined classes.

deterministic procedure. The output from the analyzer usually includes not only the decision of accepting the string generated by the given grammar, but also the derivation tree of the string, which, in turn, gives the complete description of the pattern and its subpatterns. In this chapter, several major techniques in syntax analysis are briefly described. Their applications as syntactic pattern recognizers will be demonstrated in later chapters.

Given a sentence x and a (context-free or context-free programmed) grammar G, construct a derivation of x and find a corresponding derivation tree. This is called " parsing," " recognizing," or " syntax analyzing " the sentence. Alternatively, the problem of parsing can be regarded in the following way: Given a particular sentence x and a grammar G, construct a triangle

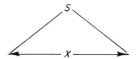

and attempt to fill the interior of the triangle with a self-consistent tree of derivations, namely, the parse. If the attempt is successful, then $x \in L(G)$; otherwise, $x \notin L(G)$. For example, consider the grammar $G = (V_N, V_T, P, S)$ where $V_N = \{S, T\}$, $V_T = \{I, +, *\}$, $I \in \{a, b, c\}$, and P:

$$S \to T \qquad T \to T * I$$
$$S \to S + T \qquad T \to I$$

The derivation tree of the sentence $x = a * b + c * a + b$ is

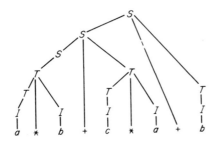

It is in principle unimportant how we attempt to fill the interior of the triangle. We may do it from the top (the root of the tree) toward the bottom (called " top-down " parsing),

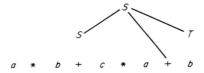

or from the bottom toward the top (called "bottom-up parsing)

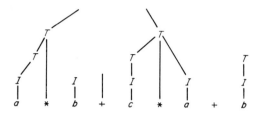

Both strategies are called "left–right," since the general order of processing the symbols in the sentence is from left to right whenever possible. Whatever the strategy used, the individual steps consist of trying to find a production to fit the locality under consideration. For example, the production

$$T \to T * I$$

is used in the last diagram. This particular step fits the local situation and will also turn out to be correct in the total parse. On the other hand, if the production

$$S \to S + T$$

is used in the local situation, as shown in the following diagram

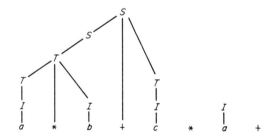

we would later have found it impossible to generate a self-consistent parse from this situation, and we would have to come back and try a different local step (backtracking). The fact that the trial attempts have to be made that later turn out to be wrong makes the parsing process potentially inefficient.

Consider the choice of the local step, regardless of the order in which we are trying to fill the triangle. A production is selected only if it passes some a priori tests. For example, a production is selected only if it fits the local situation. The a priori tests can be more elaborate. A production may be rejected because it must necessarily lead to more symbols in the terminal sequence than are available, or because it forces the use of a terminal that does not occur. If there were, in the grammar under consideration, a production

$$S \to S * T$$

it would be useless to use it if the sentence did not contain $*$. Another possible a priori test for rejecting a production depends on the first symbol of the phrase that has to be derived from it. If the desired symbol cannot be derived as the first terminal, the production should be rejected. After a production has been selected that cannot be faulted by the a priori tests, it is inserted, and we carry on to the next step. If it is found later that the resulting structure cannot be made into a proper parsing of the sentence, then we shall have to return, select another production, and try again. If it is necessary to find all the parses, then all the a priori satisfactory productions must be tried. If the language is known to be unambiguous, or if one parse is sufficient, then the process can be terminated when a satisfactory parse is found.

The speed of parsing depends on avoiding false trials, which cause work to be done that is later rejected. If the order of selecting the steps and the a priori tests is such that at each step only one production is possible, then the parsing method cannot be significantly improved (unless the a priori tests are very slow). But suppose that there are several choices for each step. Then the total number of possible parses increases exponentially. The choice of the a priori tests and the order of operation are therefore extremely important. This can be illustrated by considering the grammar $G = (V_N, V_T, P, S)$ where $V_N = \{S\}$, $V_T = \{a\}$, and P:

$$S \to aSa$$
$$S \to a$$

For the sentence $x = aaaaa$, if it is parsed from top down, the first step is

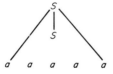

the second step is

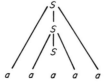

the third step is

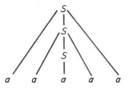

We rely on the a priori test that if the number of unattached symbols is one, then we use

$$S \rightarrow a$$

Otherwise, we use $S \rightarrow aSa$. Thus, there is only one way of proceeding.

In this chapter, parsing algorithms for context-free and context-free programmed languages are briefly presented. Several context-sensitive parsing algorithms, though recently proposed [20, 21], have not yet been tested for syntactic pattern recognition applications. Hence, context-sensitive parsing algorithms are not included in the present chapter.

4.2.   Top-Down Parsing

The top-down method starts with the sentence symbol S and makes successive substitutions for nonterminals to try to fit the sentence. A pure top-down parser is entirely "goal-oriented." The main goal is, of course, the symbol S—a "prediction" is made that the string to be analyzed is actually a sentence with respect to the given grammar G. Therefore, the first step is to see whether the string can be reduced to the right part $X_1 X_2 \cdots X_n$ of some production

$$S \rightarrow X_1 X_2 \cdots X_n$$

This is carried out as follows. For the application of the production to be valid, if X_1 is a terminal symbol, then the string must begin with this terminal. If X_1 is a nonterminal, a subgoal is established and tried: See whether some head of the string may be reduced to X_1. If this proves to be possible, X_2 is tested in the same manner; then X_3; and so on. If no match can be found for some X_i, then application of an alternative production $S \rightarrow X_1' X_2' \cdots X_m'$ is attempted.

Subgoals A, $A \in V_N$, are tested in the same manner; a production $A \rightarrow X_1 X_2 \cdots X_l$ is tested to see whether it can be applied. Thus, new subgoals are continuously being generated and attempted. If a subgoal is not met, failure is reported to the next high level, which must try another alternative production.

Left-recursion sometimes causes trouble in a left–right top-down parser; productions of the form $A_1 \to A_1\alpha$ may cause infinite loops [22]. This is because when A_1 becomes the new subgoal, the first step is to create again a new subgoal A_1. The left-recursion problem is sometimes solved by transforming the grammar or modifying the parser in some way [23]. The order in which the different right parts of productions for the same left part are tested can also play a large role here. If there is only one right part that contains left-recursion, this one should be the last one tested. However, this sequence might conflict with other ordering problems, such as testing the shortest right part last.

Example 4.1 Consider the grammar $G_1 = (V_N, V_T, P_1, S)$, where $V_N = \{S, T, I\}$, $V_T = \{a, b, c, f, g\}$, and P_1:

$$S \to T \qquad I \to a$$
$$S \to TfS \qquad I \to b$$
$$T \to IgT \qquad I \to c$$
$$T \to I$$

Let $x = afbgc$ be the sentences to be parsed or analyzed. The successive steps of the analysis are given in the following diagram, showing the remainder of the sentence, and the current parse tree at each stage. Start with S, and replace it by its alternative—in this case TfS is the correct one. In future steps, if the first symbol is a terminal, then it is compared with the first symbol of the sentence x, and if they are the same, this terminal is deleted from both the sentence and the sequence of symbols. This sequence represents the permissible structure for the rest of the sentence. If the first symbol is a nonterminal, then it is replaced by its alternatives. Since we are choosing the correct productions, the terminals will be the same; but in general they may not be, and if they are not, the parse fails.

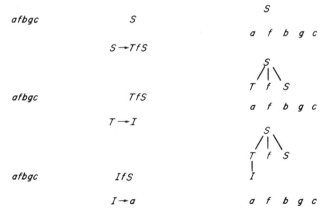

fbgc	*fS*	
bgc	*S*	
	S → T	
bgc	*T*	
	T → IgT	
bgc	*IgT*	
	I → b	
gc	*gT*	
c	*T*	
	T → I	

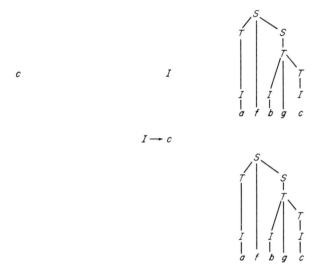

Thus, $x \in L(G_1)$ and the derivation tree of x is also given as shown in the last diagram.

Example 4.2   Consider the grammar $G_2 = (V_N, V_T, P_2, S)$, where $V_N = \{S, T, I\}$, $V_T = \{a, b, c, f, g\}$, and P_2:

$$S \to T \qquad I \to a$$
$$S \to SfT \qquad I \to b$$
$$T \to TgI \qquad I \to c$$
$$T \to I$$

It is noted that the productions $S \to SfT$ and $T \to TgI$ may cause infinite loops and the parsing process will never terminate (left-recursion). There is an a priori test that will remove the danger of looping in this fashion. Every nonterminal gives rise to a terminal sequence containing at least one terminal. Hence, if the number of symbols in the remainder parse is greater than the number of terminals left in the sentence, the parse cannot be correct. This test is useful in practice only if the length of the sentence is short.

A useful a priori test can be devised which uses the current symbol of the sentence (modified top-down parser). Suppose that the current symbol is b and that we are about to replace the leftmost nonterminal A by its alternative productions. We need only select those which can yield phrases starting with b. If initially the grammar is processed to produce a matrix of terminals against nonterminals, indicating which nonterminals can begin with which terminals, then this matrix can be used to weed out the productions that cannot give rise to the starting terminal. Such a matrix can easily be constructed. Suppose that we start with a matrix of terminals and nonterminals against nonterminals, indicating those terminals and nonterminals that begin the productions for the nonterminals. For example, consider the grammar $G = (V_N, V_T, P, S)$, where $V_N = \{T, S\}$, $V_T = \{a, b, c, d\}$, and P:

$$S \rightarrow Tb \qquad T \rightarrow Td$$
$$S \rightarrow a \qquad T \rightarrow c$$

The following matrix

	S	T	a	b	c	d
S	0	1	1	0	0	0
T	0	1	0	0	1	0

indicates that S begins with T or a, and that T begins with T or c. Look into the row for each nonterminal or the rows of the nonterminals present in that row, and repeat this until the whole matrix is stationary. Then the matrix indicates all the terminals which can begin the nonterminals:

	S	T	a	b	c	d
S	0	1	1	0	1	0
T	0	1	0	0	1	0

An alternative way of obtaining a similar effect is to take the grammar, which contains no left-recursive nonterminals, and repeatedly substitute the alternative productions for nonterminals occurring at the left of productions, until all the productions begin with a terminal (Greibach normal form). For example, consider the grammar G_1 in Example 4.1; its Greibach normal form is

$$S \rightarrow I \qquad S \rightarrow IgTfS \qquad I \in \{a, b, c\}$$
$$S \rightarrow IgT \qquad T \rightarrow I$$
$$S \rightarrow IfS \qquad T \rightarrow IgT$$

This gives, for each nonterminal–terminal pair, the productions of the nonterminal that begin with the terminal. The a priori test is to use the productions associated with the nonterminal at the start of the remainder of the parse and the terminal at the start of the remainder of the sentence. This method of analysis is referred to as predictive analysis [24]. Since the grammar in normal form may be somewhat different from the original grammar, steps may have to be taken to recover the required parse.

Shaw [25] has used a top-down parser to analyze a set of spark chamber pictures produced in high-energy particle physics experiments. In fact, Shaw's parsing algorithm is an n-dimensional analogy of a classical top-down string parser. The reasons for choosing a pure top-down goal-oriented parser are

(1)   The language portion of the analysis is conceptually very simple, consisting primarily of stepping through the pattern grammar.

(2)   The syntax directly expresses the algorithm for analysis.

(3)   Any inefficiencies due to the backtracking caused by false trials would be insignificant compared to the primitive recognition time. Once a pattern primitive is recognized, it is stored; thus, if a goal fails, its primitives may be used later in the analysis without recognition.

(4)   Goal-oriented analysis is beneficial for primitive recognition. Each primitive recognizer could operate independently with its own preprocessing and often need not be as precise as a scheme that requires a search for all primitives at any stage in the analysis. The same advantages hold over global methods that produce a list of all the primitives in a picture. These advantages are achieved because the concatenation operators in conjunction with the previously found primitives tell the parser where to look for the next primitive (Appendix B).

4.3   Bottom-Up Parsing

In contrast to the top-down method, in which by successive substitutions the sentence symbol S is expanded to produce a sentence, the bottom-up method starts with the sentence and applies the productions backward, trying to contract to the sentence symbol. In other words, the string is searched for substrings which are right parts of productions. These are then replaced by the corresponding left side. A pure bottom-up parser has essentially no long-range goals, except of course the implicit goal S. The following example illustrates a step-by-step substitution for a left–right bottom-up parsing.

Example 4.3   Consider the grammar G_1 in Example 4.1 $x = afbgc$. The successive steps of the bottom-up analysis are given in the following diagram.

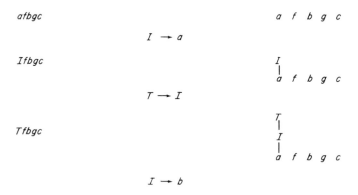

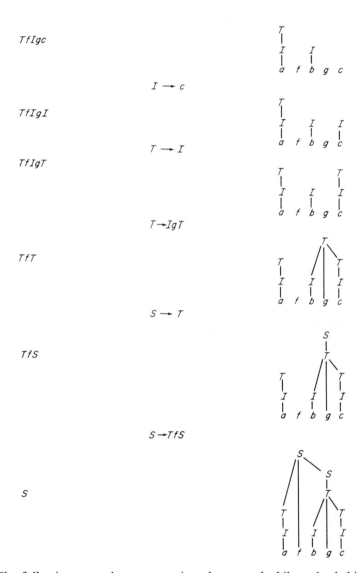

$TfIgc$

$I \rightarrow c$

$TfIgI$

$T \rightarrow I$

$TfIgT$

$T \rightarrow IgT$

TfT

$S \rightarrow T$

TfS

$S \rightarrow TfS$

S

The following procedure summarizes the general philosophy behind left–right bottom-up parsing. Suppose that we have a sentential form s and its derivation or syntax tree. Define a "phrase" of s to be the set of leaves (end nodes) of some subtree of the derivation tree. The leftmost phrase which contains no phrases other than itself is called the "handle" of s relative to the derivation tree. For example, in the derivation tree of the sentence $afbgc$ in Example 4.3 there are five phrases: a, b, c, bgc, $afbgc$. Four of these contain no other phrases: a, b, c, and bgc. The handle is the leftmost such phrase: a. Starting with a sentential form $x = s_0$, repeat the following steps for $i = 0, 1, \ldots, n$ until $s_n = S$ has been produced:

(1) Find the handle of s_i.
(2) Replace the handle of s_i by the label on the node naming the corresponding branch, to obtain the sentential form s_{i+1}.
(3) Prune the tree by deleting the handle from it. The sequence $S = s_n \Rightarrow s_{n-1} \Rightarrow \cdots \Rightarrow s_1 \Rightarrow s_0$ is then a derivation of $s_0 = x$. For example, given the sentential form $s_0 = afbgc$ and its derivation tree in Example 4.3, the handle a is replaced by I and pruned from the tree, yielding $s_1 = Ifbgc$, and

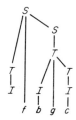

For s_1 and its derivation tree, shown above, the handle I is replaced by T and pruned from tree, yielding $s_2 = Tfbgc$, and

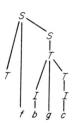

Similarly,

$S_3 = TfIgc$ $\qquad\qquad\qquad\qquad S_4 = TfIgI$

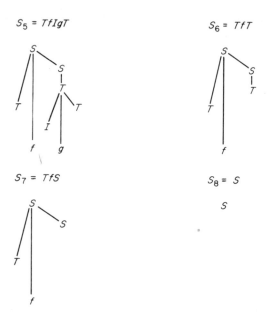

$S_5 = TfIgT$

$S_6 = TfT$

$S_7 = TfS$

$S_8 = S$

The pure bottom-up method is not particularly efficient because a large number of false trials or errors may be made. However, a powerful a priori test can be applied. Consider Example 4.3 at the stage $Ifbgc$. The only productions worth applying are those with the right side beginning with I. Each such production is part of the definition of an intermediate nonterminal. Unless that nonterminal can begin a sentence, there is no point in using the production. In this case two productions are applicable

$$T \to I$$
$$T \to IgT$$

and since T can begin S, both must be tried. The second fails, leaving $Tfbgc$. Once again, there are two applicable productions

$$S \to T$$
$$S \to TfS$$

and both must be tried. The first fails. Now we know that the parse is

and therefore that the remainder has to start with an S. Thus, it is only worth applying productions to the start of *bgc* which define nonterminals capable of starting *S*.

A matrix similar to that described in the top-down method can be formed to indicate which nonterminals can start examples of other nonterminals. It is easy to see that left-recursion presents no problem in the bottom-up parsing.† In general, two different methods are used to handle the situation of false trials. The first method is to back up (backtracking) to a point where another alternative may be tried. This procedure involves restoring parts of the string to a previous form or erasing some of the connections made on the derivation tree. The second method is to carry out all possible parses in parallel. As some of them lead to dead ends (no more reductions are possible), they are dropped. However, in practice, many parsers have been modified and become sophisticated enough so that, with certain restrictions on the grammar, backtracking or parallelism is unnecessary.

Narasimhan and Ledley *et al.* have used the bottom-up method for the analysis of bubble chamber pictures and chromosome patterns, respectively [26–28]. Details of the grammars that they constructed for bubble chamber pictures, English characters and chromosome patterns are described in Appendixes A and D.

A general comparison of the parsing methods is difficult for two reasons. First, some grammars are parsed more efficiently by top-down methods and others by bottom-up methods, so the efficiency depends on the grammar. Second, if a particular grammar is found to be parsed inefficiently by one method, a transformed version of it can usually be found which can be parsed efficiently by that method. Most parsers that produce all the parses have been based on top-down methods. Most of the special-purpose parsers for restricted kinds of grammar can be thought of as based on bottom-up methods. The existing top-down parsers therefore accept a wider class of grammars but tend to be less efficient. Griffiths and Petrick [29] have made a comparison of parsers on the basis of the efficiency of Turing machines corresponding to each of the parsers. They have found the number of steps required by a Turing machine to parse various grammars in various ways, and have also shown that there is always an equivalent grammar which can be parsed by a top-down method not much more slowly than the original was parsed by a bottom-up method.

4.4 *LR(k)* Grammars

In a pure bottom-up left–right parser, at each step there are usually many substrings that can be replaced by a nonterminal; hence, we must often try a large number of choices. For certain classes of grammars this process can be

---

† Readers may try the bottom-up method for Example 4.2.

carried out in a simple, deterministic manner. One such class of grammars is called $LR(k)$, which stands for left–right parsing with a "look-ahead" of k symbols [31]. Intuitively, we say a context-free grammar $G = (V_N, V_T, P, S)$ is $LR(k)$ if for any sentential form s the following holds: There is a unique way to write $s \Longrightarrow \beta\gamma\delta$ such that there is a rightmost derivation $S \overset{*}{\Longrightarrow} \beta A\delta \Longrightarrow \beta\gamma\delta$, A having been replaced by γ at the last step. Moreover, A and γ can be determined uniquely by scanning s from left to right up to a point at most k symbols beyond γ. To handle the situation where the k look-ahead symbols extend beyond the end of the sentence, we look at strings derivable from the string $S\$^k$ rather than S, $\$ \notin V$.

Formally, we defined that G is $LR(k)$ if the following condition holds for every string $\alpha\beta x_1 x_2$, where α, $\beta \in V^*$ and x_1, $x_2 \in V_T{}^*$, with $|x_1| = k$ such that $S\$^k \overset{*}{\underset{rt}{\Longrightarrow}} \alpha\beta x_1 x_2$.† If the next to the last step of the derivation is $\alpha A x_1 x_2$, so that

$$S\$^k \overset{*}{\underset{rt}{\Longrightarrow}} \alpha A x_1 x_2 \underset{rt}{\Longrightarrow} \alpha\beta x_1 x_2$$

and there is some other string $\alpha\beta x_1 x_3$ such that

$$S\$^k \overset{*}{\underset{rt}{\Longrightarrow}} \gamma B x \underset{rt}{\Longrightarrow} \alpha\beta x_1 x_3$$

then $\gamma = \alpha$, $A = B$, and $x = x_1 x_3$. In other words, in the rightmost derivation of two strings that agree up to k symbols beyond the point of the last replacement, the strings at the next to last step in the derivations must also agree up to k symbols beyond the point of the last replacement.

Example 4.4   Consider the grammar $G = (V_N, V_T, P, S)$, where $V_N = \{S\}$, $V_T = \{a, b\}$, and P:

$$S \rightarrow aSb$$
$$S \rightarrow Sb$$
$$S \rightarrow b$$
$$L(G) = \{a^n b^m \mid m > n\}$$

It can be shown that G is not $LR(1)$. In fact, G is not $LR(k)$ for any k. For example, let $\alpha = aa$, $\gamma = aaa$, $A = B = S$, $\beta = aSb$, $x_1 = b$, $x_2 = bb\$$, $x = bbbb\$$, and $x_3 = bbb\$$. Then

$$S\$ \overset{*}{\underset{rt}{\Longrightarrow}} \alpha A x_1 x_2 \underset{rt}{\Longrightarrow} \alpha\beta x_1 x_2 = aaaSbbbb\$$$

---

† The sign $\overset{\bullet}{\underset{rt}{\Longrightarrow}}$ indicates rightmost derivation, i.e., the rightmost nonterminal is replaced at each step of the derivation.

But it is also true that

$$S\$ \xrightarrow[\text{rt}]{*} \gamma Bx \xrightarrow[\text{rt}]{} \alpha\beta x_1 x_3 = aaaSbbbb\$.$$

Since $\gamma \neq \alpha$, G is not $LR(1)$.

Several important properties of $LR(k)$ grammars are listed here.

(1)   Every $LR(k)$ grammar is unambiguous.

(2)   The language generated by an $LR(k)$ grammar is accepted by a deterministic pushdown automaton. (The language is called a "deterministic language" [30].)

(3)   Every language accepted by a deterministic pushdown automaton is generated by some $LR(1)$ grammar. Thus a language is $LR(1)$ if and only if it is $LR(k)$ for some k.

(4)   If L_1 is accepted by a deterministic pushdown automaton, then there exists an $LR(0)$ grammar G such that $L(G) = L_1\$$.

Knuth [31] introduced two algorithms to determine if a context-free grammar is $LR(k)$ for a given k. The second algorithm also constructs the parser for an $LR(k)$ grammar in the form of a deterministic pushdown automaton. There is a class of grammars similar to $LR(k)$ grammars which can be parsed top-down deterministically while scanning the string from left to right; they are called $LL(k)$ (for leftmost derivation) or $TD(k)$ grammars [32, 33]. Various sub-families of the deterministic languages have received consideration. Among these are the precedence language [34], bounded-context languages [35], and simple deterministic languages [36].

4.5   An Efficient Top-Down Parser

An efficient context-free parsing algorithm has recently been developed by Earley [37]. Earley's algorithm bears a close relationship to Knuth's algorithm on $LR(k)$ grammars; however, Earley's algorithm does not require the grammar to be in any special form. In essence, Earley's algorithm is a top-down parser that carries along all possible parses simultaneously in such a way that like subparses can often be combined. This characteristic cuts down on duplication of effort and also avoids the left-recursion problem. An informal description of the algorithm is given in the following. It scans an input string $x = X_1 \cdots X_n$ from left to right, looking ahead some fixed number k of symbols. As each symbol X_i is scanned, a set of states Q_i is constructed which represents the condition of the parsing process at that point in the scan. Each state in the set Q_i is defined by a quadruple which represents (i) a production such that the parser is currently scanning a portion of the input string which is derived from its right side, (ii) a point in that production which shows how much of the production's right side the parser has accepted so far, (iii) a pointer back to the position in the input string at which the parser began to look for that instance of the production, and (iv) a k-symbol string which is a syntactically allowed successor to that instance of the production. This quadruple is expressed as a production, with a dot in it, followed by a string and an integer.

The following example grammar of simple arithmetic expressions is employed to illustrate the operation of the parser: $G = (V_N, V_T, P, S)$, where $V_N = \{S, T, E\}$, $V_T = \{a, +, *\}$, and P:

$$S \to T \qquad\qquad T \to T * E$$
$$S \to S + T \qquad\quad E \to a$$
$$T \to E$$

Let the string to be parsed be $x = a + a * a$. The first symbol scanned is a, and the state set Q_1 consists of the following states (excluding the k-symbol strings):

$$
\begin{array}{llll}
E \to a. & 0 & S \to T. & 0 \\
T \to E. & 0 & S \to S. + T & 0 \\
T \to T. * E & 0 & &
\end{array}
$$

Each state represents a possible parse for the beginning of the string, given that only the a has been scanned. All the states have 0 as a pointer, since all the productions represented must have begun at the beginning of the string. There will be one such state set for each position in the string. To aid in parsing, we place $(k + 1)$ \$ symbols ($\$ \notin (V_N \cup V_T)$) at the right end of the input string.

To begin the algorithm, the parser puts the single state

$$\varnothing \to .S \quad \$\$^k \quad 0$$

into state set Q_0, where $\varnothing$ is a new nonterminal. In general, the parser operates on a state set Q_i as follows. It processes the states in the set in order, performing one of three operations on each one, depending on the form of the state. These operations may add more states to Q_i and may also put states in a new state set Q_{i+1}. We describe these three operations in terms of the example above. In grammar G, with $k = 1$, Q_0 starts as the single state

(1) $\varnothing \to .S \quad \$\$ \quad 0$

(i)   The predictor operation is applicable to a state when there is a nonterminal to the right of the dot. It causes the parser to add one new state to Q_i for each alternative of that nonterminal. The dot is put at the beginning of the production in each new state since none of the symbols have yet been scanned. The pointer is set to i since the state is created in Q_i. Thus the predictor adds to Q_i all productions that might generate substrings beginning at X_{i+1}.

In the present example, the parser adds to Q_0

$$(2)\quad S \to .S + T \quad \$ \quad 0$$

$$(3)\quad S \to .T \qquad \$ \quad 0$$

The k-symbol look-ahead string is \$, since it is after S in the original state. These two states are now in processing. The predictor is also applicable to them. Applying to (2), it produces

$$(4)\quad S \to .S + T \quad + \quad 0$$

$$(5)\quad S \to .T \qquad + \quad 0$$

with a look-ahead symbol $+$, since it appears after S in (2). Applying to (3), it produces

$$(6)\quad T \to .T * E \quad \$ \quad 0$$

$$(7)\quad T \to .E \qquad \$ \quad 0$$

with a look-ahead symbol \$, since T is the last in the production and \$ is its look-ahead symbol. Now, the predictor, applying to (4), produces (4) and (5) again. Applying to (5), it produces

$$T \to .T * E \quad + \quad 0$$

$$T \to .E \qquad + \quad 0$$

The rest of Q_0 consists of

$$T \to .T * E \quad * \quad 0$$

$$T \to .E \qquad * \quad 0$$

$$E \to .a \qquad \$ \quad 0$$

$$E \to .a \qquad + \quad 0$$

$$E \to .a \qquad * \quad 0$$

The predictor is not applicable to any of the last three states. Instead, the scanner is applicable, because it is applicable just in case there is a terminal to the right of the dot.

(ii)   The scanner operation compares the terminal to the right of the dot with X_{i+1}, and if they match, it adds the state to Q_{i+1}, with the dot moved over one in the state to indicate that the terminal symbol has been scanned.

If $X_1 = a$, then Q_1 is

$$(8) \quad E \to a. \quad \$ \quad 0$$
$$E \to a. \quad + \quad 0$$
$$E \to a. \quad * \quad 0$$

These states are added by the scanner. If the parser finishes processing Q_i and Q_{i+1} remains empty, an error has occurred in the input string. Otherwise, it starts to process Q_{i+1}.

(iii)   The completer operation is applicable to a state if its dot is at the end of its production. Thus, in the present example, it is applicable to each of these states in Q_1. It compares the look-ahead string with $X_{i+1} \cdots X_{i+k}$. If they match, it goes back to the state set indicated by the pointer, in this case Q_0, and adds all states from Q_0 which have E to the right of the dot. It moves the dot over E in these states. If $X_2 = +$, then the completer applying to (8) produces

$$(9) \quad T \to E. \quad \$ \quad 0$$
$$T \to E. \quad + \quad 0$$
$$T \to E. \quad * \quad 0$$

Applying the completer to the second state in (9) produces

$$(10) \quad S \to T. \qquad \$ \quad 0$$
$$S \to T. \qquad + \quad 0$$
$$T \to T. * E \quad \$ \quad 0$$
$$T \to T. * E \quad + \quad 0$$
$$T \to T. * E \quad * \quad 0$$

From the second state in (10), we get

$$(11) \quad \varnothing \to S. \$ \qquad \$ \quad 0$$
$$S \to S. + T \quad \$ \quad 0$$
$$S \to S. + T \quad + \quad 0$$

The scanner then adds to Q_2

$$(12) \quad S \to S + .T \quad \$ \quad 0$$
$$S \to S + .T \quad + \quad 0$$

If the algorithm ever produces a Q_{i+1} consisting of the single state

$$\varnothing \to S\$. \quad \$ \quad 0$$

then the parser has correctly scanned an S and the $\$$. So we have finished with the string, and it is accepted as a sentence generated by G.

In order to implement the algorithm efficiently, a look-ahead of $k = 1$ is recommended. The algorithm can handle a larger class of context-free grammars in linear time than most of the restricted algorithms (using quadric or cubic time bound). A step-by-step operation of the parser for the example above is given in the following, where $k = 1$.

```
Q0:      ∅ → .S$      $   0      T → .T * E    *   0
(X1 = a) S → .S + T   $   0      T → .E        $   0
         S → .S + T   +   0      T → .E        +   0
         S → .T       $   0      T → .E        *   0
         S → .T       +   0      E → .a        $   0
         T → .T * E   $   0      E → .a        +   0
         T → .T * E   +   0      E → .a        *   0


Q1:      E → a.       $   0      S → T.        +   0
(X2 = +) E → a.       +   0      T → T. * E    $   0
         E → a.       *   0      T → T. * E    +   0
         T → E.       $   0      T → T. * E    *   0
         T → E.       +   0      ∅ → S. $      $   0
         T → E.       *   0      S → S. + T    $   0
         S → T.       $   0      S → S. + T    +   0


Q2:      S → S + .T   $   0      T → .E        $   2
(X3 = a) S → S + .T   +   0      T → .E        +   2
         T → .T * E   $   2      T → .E        *   2
                                 E → .a        $   2
         T → .T * E   +   2      E → .a        +   2
         T → .T * E   *   2      E → .a        *   2


Q3:      E → a.       $   2      S → S + T.    $   0
(X4 = *) E → a.       +   2      S → S + T.    +   0
         E → a.       *   2      T → T. * E    $   2
         T → E.       $   2      T → T. * E    +   2
         T → E.       +   2      T → T. * E    *   2
         T → E.       *   2


Q4:      T → T * .E   $   2      E → .a        $   4
(X5 = a) T → T * .E   +   2      E → .a        +   4
         T → T * .E   *   2      E → .a        *   4
```

Q_5:	$E \to a.$	\$	4	$S \to S + T.$	$+$	0
$(X_6 = \$)$	$E \to a.$	$+$	4	$T \to T. * E$	\$	2
	$E \to a.$	$*$	4	$T \to T. * E$	$+$	2
	$T \to T * E.$	\$	2	$T \to T. * E$	$*$	2
	$T \to T * E.$	$+$	2	$\varnothing \to S. \$$	\$	0
	$T \to T * E.$	$*$	2	$S \to S. + T$	\$	0
	$S \to S + T.$	\$	0	$S \to S. + T$	$+$	0

Q_6: $\varnothing \to S\$.$ \$ 0

4.6   Operator Precedence Grammars

A grammar is said to be an operator grammar if there is no production of the form

$$A \to \alpha BC\beta$$

where $\alpha, \beta \in V^*$, and $A, B, C \in V_N$. This means that no sentential form contains two adjacent nonterminals.† Given an operator grammar, let s be a sentential form. We define a " prime phrase " of s to be a phrase that contains no phrase other than itself but at least one terminal. Equivalently, α is a prime phrase of at least one sentential form s if and only if α contains at least one terminal and there exists either a production $A \to \alpha$ or a production $A \to \alpha'$ where $\alpha' \overset{*}{\Longrightarrow} \alpha$ and the only productions applied in the derivation $\alpha' \overset{*}{\Longrightarrow} \alpha$ are of the form $A_i \to A_j$. The parser to be described reduces at each step the leftmost prime phrase, and not the handle. However, we still call this a bottom-up left–right parser, since it is proceeding essentially in a left to right manner.

Let us define the following three relations between terminals t_1 and t_2 of an operator grammar.

(1) $t_1 \doteq t_2$ if there is a production $A \to \alpha t_1 t_2 \beta$ or $A \to \alpha t_1 B t_2 \beta$ where $\alpha, \beta \in V^*$, and $A, B \in V_N$.

(2) $t_1 \gtrdot t_2$ if there is a production $A \to \alpha B t_2 \beta$ and a derivation $B \overset{*}{\Longrightarrow} \gamma t_1$ or $B \overset{*}{\Longrightarrow} \gamma t_1 C$ for some $\gamma \in V^*$ and $C \in V_N$.

(3) $t_1 \lessdot t_2$ if there is a production $A \to \alpha t_1 B \beta$ and a derivation

$$B \overset{*}{\Longrightarrow} t_2 \gamma \quad \text{or} \quad B \overset{*}{\Longrightarrow} C t_2 \gamma \qquad \text{for some } \gamma \text{ and } C$$

† This is not a serious restriction. Many useful grammars are operator grammars, or can easily be changed into this form.

If at most one relation holds between any ordered pair t_1, t_2 of terminals, then the grammar is called an "operator precedence grammar" and the language generated an "operator precedence language." The relations $>$, $\doteq$, and $<$ can be kept in a matrix form, called a "precedence matrix."

In an operator precedence language, these unique relations are clearly helpful when parsing is being done. For suppose that t_1 and t_2 are two neighboring terminals in a sentential form or a partly reduced sentence; that is, suppose that they are separated only by nonterminals and not by other terminals. Then they belong to the same prime phrase if and only if $t_1 \doteq t_2$. If $t_1 > t_2$, then t_1 belongs to a phrase to which t_2 does not belong, and the nonterminals between t_1 and t_2 also belong to that phrase. Similarly, if $t_1 < t_2$, then t_2 belongs to a phrase to which t_1 does not belong. Hence it is easy to determine the phrase structure of the sentence. For example, suppose $t_0 \alpha t$ is a substring of a sentential form $s = \alpha_1 t_0 \alpha t \alpha_2$ and that the terminals in α are, in order, $t_1, t_2, \ldots, t_n$ ($n \geq 1$). Now if the following relations hold between $t_0, t_1, \ldots, t_n$ and t,

$$t_0 < t \doteq t_2 \doteq \ldots \doteq t_n > t$$

then α is a prime phrase. Furthermore, the reduction of α to some A may always be executed to yield the sentential form $\alpha_1 t_0 A t \alpha_2$.†

Example 4.5   Consider the grammar $G = (V_N, V_T, P, S)$, where $V_N = \{T, P, S\}$, $V_T = \{I, f, g, (,)\}$, $I \in \{a, b, c, d\}$, and P:

$$S \rightarrow SfT \qquad T \rightarrow P$$
$$S \rightarrow T \qquad P \rightarrow (S)$$
$$T \rightarrow TgP \qquad P \rightarrow I$$

The precedence matrix is

	f	g	$($	$)$	I
f	$>$	$<$	$<$	$>$	$<$
g	$>$	$>$	$<$	$>$	$<$
$($	$<$	$<$	$<$	$\doteq$	$<$
$)$	$>$	$>$	$\doteq$	$>$	
I	$>$	$>$		$>$	

We will now parse

$$a \; f \; b \; g \; (\; c \; f \; d \;)$$

using N to stand for unknown nonterminals, and showing the formation of

---

† It should be noted that, in terms of the concept of $LR(k)$ grammars, an operator precedence grammar is an $LR(0)$ grammar.

the tree of phrases:

$$a \quad f \quad b \quad g \quad (\quad c \quad f \quad c \quad)$$
$$\gtrdot \quad \lessdot \quad \gtrdot \quad \lessdot \quad \lessdot \quad \gtrdot \quad \lessdot \quad \gtrdot$$

$$N \quad f \quad b \quad g \quad (\quad c \quad f \quad d \quad)$$
$$\lessdot \quad \gtrdot \quad \lessdot \quad \lessdot \quad \gtrdot \quad \lessdot \quad \gtrdot$$

$$N \quad f \quad N \quad g \quad (\quad c \quad f \quad d \quad)$$
$$\lessdot† \quad \lessdot \quad \lessdot \quad \gtrdot \quad \lessdot \quad \gtrdot$$

$$N \quad f \quad N \quad g \quad (\quad N \quad f \quad d \quad)$$
$$\lessdot \quad \lessdot \quad \lessdot \quad \lessdot \quad \gtrdot$$

$$N \quad f \quad N \quad g \quad (\quad N \quad f \quad N \quad)$$
$$\lessdot \quad \lessdot \quad \lessdot \quad \gtrdot$$

$$N \quad f \quad N \quad g \quad (\quad N \quad)$$
$$\lessdot \quad \lessdot \quad \doteq$$

$$N \quad f \quad N \quad g \quad N$$
$$\lessdot$$

$$N \quad f \quad N$$
$$\lessdot$$

$$N$$

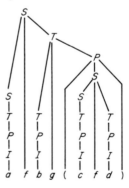

The fact that the parse is really

† The relation is between f and g; the N is ignored.

has to be deduced by adding the effect of productions like

$$S \to T$$

and by finding the correct nonterminals with which to label the nodes. A flow diagram for the parser using operator precedences is shown in Fig. 4.2.

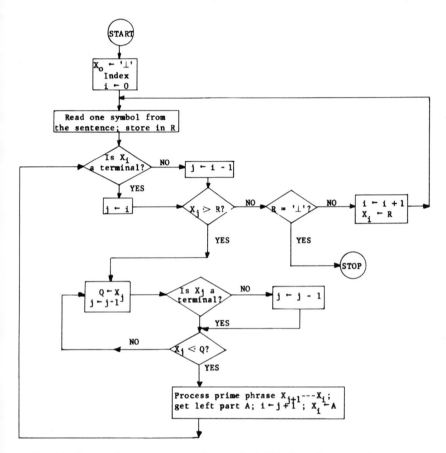

Fig. 4.2. Parser using operator precedences. The ' $\perp$ ' is the end marker of the sentence; that is, the sentence to be analyzed is $\perp x \perp$. $X_0, X_1, \ldots, X_i$ is a stack holding a position of the sentential form under analysis.

The precedence relations between terminals can often be expressed by two weights for each terminal. Consequently, the space needed for the relations may be reduced to two vectors if two integer "precedence functions" $f(t)$ and $g(t)$ can be found such that $t_1 \lessdot t_2$ implies $f(t_1) < g(t_2)$. $t_1 \doteq t_2$ implies $f(t_1) = g(t_2)$, and $t_1 \gtrdot t_2$ implies $f(t_1) > g(t_2)$. For example, the precedence functions for the language specified in Example 4.5 are

t	$f(t)$	$g(t)$
f	3	2
g	5	4
(	1	6
)	6	1
I	5	6

Floyd [34] gives algorithms for processing a grammar to determine if it is an operator precedence grammar, to find the precedence matrix, and to produce the precedence functions $f(t)$ and $g(t)$ if they exist. It is rather difficult to figure out a good error recovery scheme if the precedence functions are used, since an error or a false trial can be detected only when a probable prime phrase turns out not to be one. With the full matrix, an error is detected whenever no relation exists between the top terminal stack symbol and the incoming symbol. Therefore the functions should be used only if a previous pass has provided a complete syntax check.

4.7   Precedence and Extended Precedence Grammars

Wirth and Weber [38] modified Floyd's precedence concept. The grammar is not restricted to an operator grammar and the precedence relations $<$, $\doteq$, and $>$ may hold between all pairs of symbols, including terminals and nonterminals:

(1) $X_1 \doteq X_2$ if there is a production $A \to \alpha X_1 X_2 \beta$.

(2) $X_1 > X_2$ if there is a production $A \to \alpha B X_2 \beta$ (or $A \to \alpha B C \beta$) and a derivation $B \overset{*}{\Longrightarrow} \gamma X_1$ (and $C \overset{*}{\Longrightarrow} X_2 \delta$) for some γ, δ.

(3) $X_1 < X_2$ if there is a production $A \to \alpha X_1 B \beta$ and a derivation $B \overset{*}{\Longrightarrow} X_2 \gamma$ for some γ.

If at most one relation holds between any pair of symbols X_1 and X_2, and if no two productions have identical right parts, then the grammar is called a "precedence grammar" and the language a "precedence language." Any sentence of a precedence language has a unique syntax tree. When parsing, as long as either the relation $<$ or $\doteq$ holds between the top stack symbol X_i and the incoming symbol t, t is pushed into the stack. When $X_i > t$, then the stack is searched downward for the configuration

$$X_{j-1} < X_j \doteq \ldots \doteq X_{i-1} \doteq X_i$$

The handle is then $X_j \cdots X_i$ and is replaced by the left part A of the unique production $A \to X_j \cdots X_i$. The main difference between this technique and that discussed in the previous section is that the precedence relations may hold between any two symbols and not just between terminals; therefore, the handle and not the prime phrase is reduced. Algorithms for determining the precedence matrix and precedence functions similar to Floyd's are given by Wirth and Weber [38]. However, it is usually necessary to transform the grammar in order to avoid some conflicts of precedence relations. It was demonstrated by Fischer [39] and Learner and Lim [40] that every context-free grammar G_1 can be converted into a Wirth–Weber precedence grammar G_2 where $L(G_1) = L(G_2)$. The following example illustrates the parsing process using Wirth–Weber precedences.

Example 4.6 Consider the grammar of Example 4.5. It has been found that $f \doteq T$ from $S \to SfT$, and that $f < T$ from $S \to SfT$ and $T \to TgP$. There is also a conflict in that $(\doteq S$ and $(<S$ both hold. The grammar can be modified by adding extra nonterminals

$$
\begin{array}{ll}
S \to U & V \to VgP \\
U \to UfT & V \to P \\
U \to T & P \to (S) \\
T \to V & P \to I
\end{array}
$$

thus giving a precedence grammar. A step-by-step parsing is now demonstrated

```
a  f  b  g  (  c  f  d  )            U  f  V  g  (  U  f  d  )
   >                                    ≐  <  ≐  <  <  ≐  <  >
P  f  b  g  (  c  f  d  )            U  f  V  g  (  U  f  P  )
   >                                    ≐  <  ≐  <  <  ≐  <  >
V  f  b  g  (  c  f  d  )            U  f  V  g  (  U  f  V  )
   >                                    ≐  <  ≐  <  <  ≐  <  >
T  f  b  g  (  c  f  d  )            U  f  V  g  (  U  f  T  )
   >                                    ≐  <  ≐  <  <  ≐  ≐  >
U  f  b  g  (  c  f  d  )            U  f  V  g  (  U  )
   ≐  <  >                              ≐  <  ≐  <  <  >
U  f  P  g  (  c  f  d  )            U  f  V  g  (  S  )
   ≐  <  >                              ≐  <  ≐  <  ≐  ≐
U  f  V  g  (  c  f  d  )            U  f  V  g  P
   ≐  <  ≐  <  <  <                      ≐  <  ≐  ≐
U  f  V  g  (  P  f  d  )            U  f  V
   ≐  <  ≐  <  <  >                      ≐  <
U  f  V  g  (  V  f  d  )            U  f  T
   ≐  <  ≐  <  <  >                      ≐  ≐
U  f  V  g  (  T  f  d  )            U
   ≐  <  ≐  <  <  >                   S
```

McKeeman [41] extended Wirth and Weber's concept by first separating the precedence matrix into two matrices, one for looking for the tail of the handle, the other for looking for the head of a handle, and then having the parser look at more context so that fewer precedence conflicts arise. The parser thus constructed will therefore accept a much wider class of grammars ("extended precedence grammars").

Anderson and Lee have used the Wirth–Weber precedence analysis for the recognition of hand-printed two-dimensional mathematical expressions and chromosome patterns, respectively [42, 43]. Chang has extended the operator precedence to the two-dimensional case by considering both operator precedence and operator dominance. The two-dimensional precedence analysis has been applied to the recognition of two-dimensional mathematical expressions and the format of printed material [44].

4.8   Syntax Analysis of Context-Free Programmed Languages

This section presents a method for analyzing strings to determine whether they belong to the language generated by a given context-free programmed grammar [45]. The analyzer is a top-down parsing algorithm consisting of a generation algorithm and a backtracking algorithm, which together can systematically trace out the generation or derivation tree associated with the given context-free programmed grammar. In the process, every valid generation of a string to be analyzed is found and reported. The basic concept is similar to the idea suggested by Chartres and Florentin [46] for the analysis of context-free languages. The only restriction on the context-free programmed grammar is that it contains no cycles of nonterminals that do not increase the length of the string being generated (e.g. $A \xrightarrow{r_1} B$, $B \xrightarrow{r_2} C$, $C \xrightarrow{r_3} A$, followed by another application of r_1). Practically speaking, this is a very minor restriction, but it guarantees that the parsing process will always terminate. It ensures that every path down the derivation tree produces a progressively longer sentential form.† Since any candidate analysis is terminated if the length of the current sentential form exceeds the length of the string being analyzed, and since there are at most a finite number of such candidate analyses, the procedure must eventually halt.

To implement the backtracking algorithm, the grammar is modified by the parser in a manner that, in effect, embeds the context-free programmed language in a bracket language [47]. In the case of context-free grammars, a bracket language has the feature that every string contains within itself the complete specification of the way in which it was generated. In the case of context-free programmed grammars, each string contains a history of the

---

† By definition, no production of a context-free programmed grammar may generate the empty string.

successful productions used to generate it. When the parser is backtracking over a step that involves the successful application of a production, the bracket configuration indicates which portion of the sentential form was produced by that step and must therefore be reduced to a single nonterminal.

Example 4.7   Consider the context-free programmed grammar $G_p = (V_N, V_T, J, P, A)$, where $V_N = \{A, B, C\}$, $V_T = \{a, b, c\}$, $J = \{1, 2, 3, 4, 5\}$. and

Label	Core	$S(U)$	$F(W)$
1	$A \to aBC$	$S(\{2,4\})$	$F(\varnothing)$
2	$B \to aBB$	$S(\{3\})$	$F(\varnothing)$
3	$C \to CC$	$S(\{2,4\})$	$F(\varnothing)$
4	$B \to b$	$S(\{4\})$	$F(\{5\})$
5	$C \to c$	$S(\{5\})$	$F(\varnothing)$

$$L(G_p) = \{a^n b^n c^n \mid n \geq 1\}$$

Figures 4.3 is a schematic diagram of the analysis of the string *abc* with respect to this context free programmed grammar. The notations used are

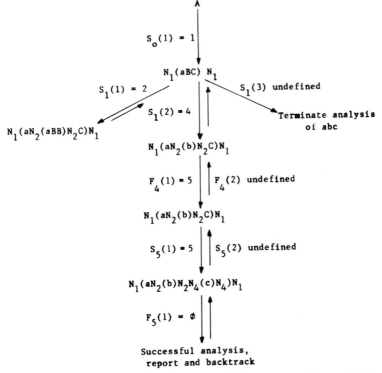

Fig. 4.3. Analysis of the string *abc* in terms of the grammar of Example 4.7.

explained as follows.

(1)   For any nonterminal $N \in V_N$, $N_i(\gamma)N_i$ indicates that the string γ was generated from the nonterminal N which was rewritten as γ at the ith step.

(2)   A downward arrow indicates a generative step; an upward arrow indicates backtracking.

(3)   The branch labels have the form $\xi_i(j) = r_k$, where ξ is S (success) or F (failure); subscript i indicates the application of the ith production; j indicates the selection of the jth branch in the ξ field, which is the branch to

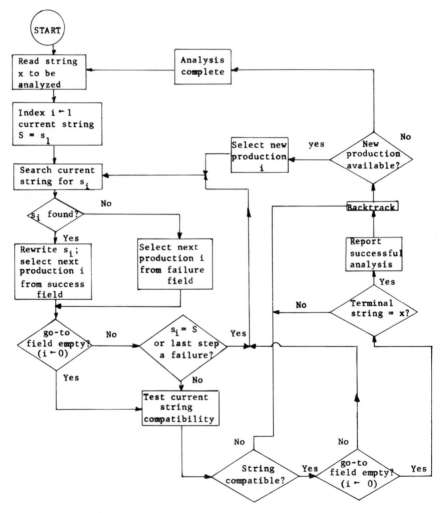

Fig. 4.4. Parser for context-free programmed grammars.

the production labeled r_k. By definition, the first (topmost) label is $S_0(1) = 1$; $\xi_i(j) = \varnothing$ indicates the termination of a path down the tree, which constitutes a successful parsing if the current sentential form is identical to the string being analyzed. The analysis of the string is complete when $\xi_i(j)$ is undefined (or empty) for $i = 1$ and some j.

Notice that the parser must continue even after a successful analysis is produced, since, in general, it is not known a priori whether the grammar is unambiguous and it is desired to produce all possible parses. The flow diagram of the parser is given in Fig. 4.4.

Swain and Fu [48] have applied the analysis to the recognition of "noisy squares" which are described by a language generated by a stochastic context-free programmed grammar.

References

1. R. W. Floyd, the syntax of programming languages—A survey. *IEEE Trans. Electron. Comput.* **EC-13**, 346–353 (1964).
2. J. Feldman and D. Gries, Translater writing systems. *Comm. ACM* **11**, 77–113 (1968).
3. J. M. Foster, *Automatic Syntactic Analysis*. Amer. Elsevier, New York, 1970.
4. F. R. A. Hopgood, *Compiling Techniques*. Amer. Elsevier, New York, 1969.
5. D. G. Hays, ed., *Readings in Automatic Language Processing*. Amer. Elsevier, New York, 1966.
6. T. E. Cheatham and K. Sattley, Syntax directed compiling. *Proc. AFIPS S   Joint Comput. Conf., 1964*, **25**, pp. 31–57.
7. S. Rosen, ed., *Programming Systems and Languages*. McGraw-Hill, New York, 1967.
8. R. B. Banerji, Some studies in syntax-directed parsing. In *Computation in Linguistics* (P. L. Garvin and B. Spolsky, eds.). Indiana Univ. Press, Bloomington, 1966.
9. R. L. Grimsdale, F. H. Summer, C. J. Tunis, and T. Kilburn, A system for the automatic recognition of patterns. *Proc. IEEE* **108B**, 210–221. (1959).
10. P. J. Knoke and R. G. Wiley, A linguistic approach to mechanical pattern recognition. *IEEE Comput. Conf., September 1967*, pp. 142–144, Conf. Rec.
11. Pavlidis, Analysis of set pattern. *Pattern Recognition* **1** (1968).
12. R. Pavlidis, Representation of figures by labeled graphs. *Pattern Recognition* **4**, 5–16 (1972).
13. T. Pavlidis, Structural pattern recognition: Primitives and juxtaposition relations. In *Frontiers of Pattern Recognition* (S. Watanabe, ed.). Academic Press, New York, 1972.
14. J. Myloponlos, A graph pattern matching operation. *Proc. Annu. Conf. Inform. Sci. and Syst., 6th, Princeton, 1972*, pp. 330–336.
15. E. H. Sussenguth, A graph-theoretic algorithm for matching chemical structures. *J. Chem. Doc.* **5**, 36–43. (1965).
16. P. H. Winston, *Learning Structural Descriptions From Examples*. Rep. MAC-TR-76, Project. MIT, Cambridge, Massachusetts, 1970.
17. A. Guzman, Analysis of curved line drawings using context and global information. *Machine Intelligence 6* (B. Meltzer and D. Michie, eds.), pp. 325–376. Edinburgh Univ. Press, Edinburgh, 1971.
18. R. O. Duda and P. E. Hart, Experiments in scene analysis. *Proc. Nat. Symp. Ind. Robots, 1st, Chicago, Illinois, April 2–3, 1970.*

19. H. G. Barrow, A. P. Ambler, and R. M. Burstall, Some techniques for recognizing structures in pictures. In *Frontiers of Pattern Recognition* (S. Watanabe, ed.). Academic Press, New York, 1972.

20. S. Kuno, A context recognition procedure. *Math. Linguistics and Automatic Translation*, Rep. NSF-18. Computation Lab., Harvard Univ. Cambridge, Massachusetts, 1967.

21. W. A. Woods, Context-sensitive parsing. *Comm. ACM*, **13**, 437–445 (1970).

22. R. Kurki-Suonio, On top-to-bottom recognition and left recursion. *Comm. ACM* **9**, 527 (1966).

23. S. Greibach, Formal parsing systems. *Comm. ACM* **7**, 499 (1964).

24. S. Kuno, The predictive analyzer and a path elimination technique. *Comm. ACM* **8**, 453 (1965).

25. A. C. Shaw, Parsing of graph-representable pictures. *J. ACM* **17**, 453–481 (1970).

26. R. Narasimhan, Syntax-directed interpretation of classes of pictures. *Comm. ACM* **9**, 166–173 (1966).

27. R. S. Ledley *et al.*, FIDAC: Film input to digital automatic computer and associated syntax-directed pattern recognition programming system. In *Optical and Electro-Optical Information Processing System* (J. Tippet *et al.*, eds.). MIT Press, Cambridge, Massachusetts, 1965.

28. R. S. Ledley, *Programming and Utilizing Digital Computers*, Chapter 8. McGraw-Hill, New York, 1962.

29. T. V. Griffiths and S. R. Petrick, On the relative efficiencies of context-free grammar recognizers. *Comm. ACM* **8**, 289–299 (1965).

30. S. Ginsburg and S. A. Greibach, Deterministic context-free languages. *Information and Control* **9**, 620–648 (1966).

31. D. E. Knuth, On the translation of languages from left to right. *Information and Control* **8**, 607–639 (1965).

32. P. M. Lewis, II and R. E. Streans, Syntax directed-transduction. *Journal Assoc. Comput. Mach.* **15**, 464–488 (1968).

33. D. J. Rosenkrantz and R. E. Stearns, Properties of deterministic top-down grammars. *Information and Control* **17**, 226–256 (1970).

34. R. W. Floyd, Syntactic analysis and operator precedence. *J. Assoc. Comput. Mach.* **10**, 316–333 (1963).

35. R. W. Floyd, Bounded context syntactic analysis. *Comm. ACM* **7**, 62–67 (1964).

36. A. J. Korenjak and J. E. Hopcroft, Simple deterministic languages. *IEEE Annu. Symp. Switching and Automata Theory, 7th, 1966, Berkeley*, California, pp. 36–46.

37. J. Earley, An efficient context-free parsing algorithm. *Comm. ACM* **13**, 94–102 (1970).

38. N. Wirth and H. Weber, EULER-A generalization of ALGOL, and its formal definition. *Comm. ACM* **9**, Pt. I, II, 13–25, 89–99 (1966).

39. M. Fischer, Some properties of precedence languages. *Proc. ACM Symp. Theory Comput. 1st, 1969*, pp. 181–190.

40. A. Learner and A. L. Lim, Note on transformating context-free grammars to Wirth Weber precedence form. *Comput. J.* **13**, 142–144 (1970).

41. W. M. McKeeman, *An Approach to Computer Language Design*. Tech. Rep. CS 48. Comput. Sci. Dept., Stanford Univ., Stanford, California, August 1966.

42. R. H. Anderson, *Syntax-Directed Recognition of Hand-printed Two-Dimensional Mathematics*. Ph.D. Thesis, Harvard Univ., Cambridge, Massachusetts, 1968.

43. H. C. Lee and K. S. Fu, A stochastic syntax analysis procedure and its application to pattern classification. *IEEE Trans. Computers* **C-21** 660–666 (1972).

44. S. K. Chang, A method for the structural analysis of two-dimensional mathematical expressions. *Information Sci.* **2**, 253–272 (1970).

45. P. H. Swain and K. S. Fu, *Nonparametric and Linguistic Approaches to Pattern Recognition.* Tech. Rep. TR-EE 70-20. School of Elec. Eng., Purdue Univ., Lafayette, Indiana, 1970.

46. B. A. Chartres and J. J. Florentin, A universal syntax-directed top-down analyzer. *J. Assoc. Comput. Mach.* **15**, 447–464 (1968).

47. N. Chomsky and M. P. Schutzenberger, The algebraic theory of context-free languages. In *Computer Programming and Formal Systems* (P. Braffort and D. Hirschberg, eds.). North-Holland Publ., Amsterdam, 1963.

48. P. H. Swain and K. S. Fu, Stochastic programmed grammars for syntactic pattern recognition. *PATTERN RECOGNITION* **4**. 83–100 (1972).

49. S. Kuno and A. G. Oettinger, Multiple-path syntactic analyzer. *Inform. Process. Proc. IFIP (Int. Fed. Inform. Proc.) Congr. 1962*, pp. 306–311. North-Holland Publ., Amsterdam, 1962.

50. T. V. Griffiths, Turing machine recognizers for general rewriting systems. *IEEE Symp. Switchinp Circuit Theory and Lopic Desipn*, Princeton, New Jersey, *1964*, pp. 47–56, 3onf. Rec.

51. A. V. Aho and J. D. Ullman, *The Theory of Parsing, Translation, and Compiling*, Vol. I, *Parsing*. Prentice-Hall, Englewood Cliffs, New Jersey, 1972.

52. J. Feder and H. Freeman, Digital curve matching using a contour correlation algorithm. *IEEE Int. Conv. Rec.* Pt. 3 (1966).

53. G. Salton and E. H. Sussenguth, Some flexible information retrieval systems using structure matching procedures. *Proc. AFIPS Spring Joint Comput. Conf. 1964*, pp. 587–597. Spartan Books, Inc., Baltimore, Maryland, 1964.

54. A. D. Inselberg, *SAP: A Model for the Syntactic Analysis of Pictures.* Tech. Rep. Sever Inst. of Technol., Washington Univ., St. Louis, Missouri, 1968.

55. T. G. Williams, *On-Line Parsing of Hand-Printed Mathematical Expressions: Final Report for Phase II.* Rep. NASA CR-1455. Syst. Develop. Corp., Santa Monica, California, 1969.

56. A. Rosenfeld and J. W. Snively, Jr., Bounds on the complexity of grammars. In *Frontiers of Pattern Recognition* (S. Watanabe, ed.). Academic Press, New York, 1972.

57. D. A. Huffman, Impossible objects as nonsense sentences. *Machine Intelligence 6.* (B. Meltzer and D. Michie, eds.), Edinburgh Univ. Press, Edinburgh, 1971.

Chapter 5

Stochastic Languages and Stochastic Syntax Analysis

5.1 Basic Formulations

In previous chapters, languages have been used for pattern description and algorithms of syntactic analysis have been used as recognition procedures. In some practical applications, a certain amount of uncertainty exists in the process under study. For example, noise or distortion occurs in the communication, storage, or retrieval of information, or in the measurement or processing of physical patterns. Usually, an extensive preprocessing is necessary before the pattern primitives can be easily extracted from the raw data. Languages used to describe the noisy and distorted patterns under consideration are often ambiguous in the sense that one string (a pattern) can be generated by more than one grammar which specifies the patterns (strings) generated from a particular pattern class. In decision-theoretic pattern recognition terminology, this is the case of overlapping pattern classes; patterns belonging to different classes may have the same descriptions or measurement values (but with different probabilities of occurrence). In these situations, in order to model the process more realistically, the approach of using stochastic languages has been suggested [1–8]. For every string x in a language L a probability $p(x)$ can be assigned such that $0 < p(x) \leq 1$ and $\sum_{x \in L} p(x) = 1$. Then the function $p(x)$ is a probability measure defined over L, and it can be used to characterize the uncertainty and randomness of L. Also, if a pattern grammar is heuristically constructed, it may generate some "unwanted" strings (strings that do not represent patterns in the class). In this case, very small probabilities could be assigned to the unwanted strings in order to use the grammar effectively.

Two natural ways of extending the concept of formal languages to stochastic languages are to randomize the productions of grammars and the state transitions of recognition devices (acceptors), respectively. In this chapter, the major results obtained so far in stochastic grammars and stochastic syntax analysis are presented.

Definition 5.1 A stochastic phrase structure grammar (or simply stochastic grammar) is a four-tuple $G_s = (V_N, V_T, P_s, S)$ where V_N and V_T are finite sets of nonterminals and terminals; $S \in V_N$ is the start symbol;† P_s is a finite set of stochastic productions each of which is of the form

$$\alpha_i \xrightarrow{p_{ij}} \beta_{ij}, \qquad j = 1, \ldots, n_i, \qquad i = 1, \ldots, k \qquad (5.1)$$

where

$$\alpha_i \in (V_N \cup V_T)^* V_N (V_N \cup V_T)^*, \qquad \beta_{ij} \in (V_N \cup V_T)^*$$

and p_{ij} is the probability associated with the application of this stochastic production,

$$0 < p_{ij} \leq 1 \quad \text{and} \quad \sum_{j=1}^{n_i} p_{ij} = 1‡ \qquad (5.2)$$

Suppose that $\alpha_i \xrightarrow{p_{ij}} \beta_{ij}$ is in P_s. Then the string $\xi = \gamma_1 \alpha_i \gamma_2$ may be replaced by $\eta = \gamma_1 \beta_{ij} \gamma_2$ with the probability p_{ij}. We shall denote this derivation by

$$\xi \xRightarrow{p_{ij}} \eta$$

and we say that ξ directly generates η with probability p_{ij}. If there exists a sequence of strings $\omega_1, \ldots, \omega_{n+1}$ such that

$$\xi = \omega_1, \quad \eta = \omega_{n+1}, \quad \omega_i \xRightarrow{p_i} \omega_{i+1}, \qquad i = 1, \ldots, n,$$

then we say that ξ generates η with probability $p = \prod_{i=1}^{n} p_i$ and denote this derivation by

$$\xi \xRightarrow[*]{p} \eta$$

† In a more general formulation, instead of a single start symbol, a start symbol (probability) distribution can be used.

‡ If these two conditions are not satisfied, p_{ij} will be denoted as the weight w_{ij} and P_s the set of weighted productions. Consequently, the grammar and the language generated are called weighted grammar and weighted language, respectively [9].

The probability associated with this derivation is equal to the product of the probabilities associated with the sequence of stochastic productions used in the derivation. It is clear that $\xLongrightarrow[*]{p}$ is the reflexive and transitive closure of the relation $\xLongrightarrow{p}$.

The stochastic language generated by G_s is

$$L(G_s) = \left\{ (x, p(x)) \,|\, x \in V_T^*, \; S \xLongrightarrow[*]{p_j} x, \quad j = 1, \ldots, k, \quad \text{and} \quad p(x) = \sum_{j=1}^{k} p_j \right\}$$

(5.3)

where k is the number of all distinctively different derivations of x from S and p_j is the probability associated with the jth distinctive derivation of x.

Since the productions of a stochastic grammar are exactly the same as those of the nonrandomized grammar except for the assignment of the probability distribution, the set of languages generated by a stochastic grammar is the same as that generated by the nonrandomized version.

In general, a stochastic language $L(G_s)$ is characterized by (L, p) where L is a language and p is a probability distribution defined over L. The language L of a stochastic language $L(G_s) = (L, p)$ is called the characteristic language of $L(G_s)$.

Definition 5.2 For any stochastic grammar G_s, the corresponding characteristic grammar, denoted by $\overline{G}_s$, is the grammar obtained from G_s by deleting the probabilities associated with each stochastic production in G_s. Let $P_s = (P, D)$ where D is the probability measure over the production set P; then $\overline{G}_s = (V_N, V_T, P, S)$, which is obviously nonstochastic.

Based on the concept of characteristic grammars, we shall say that a stochastic grammar G_s is a stochastic type 0 (unrestricted), type 1 (context-sensitive), type 2 (context-free), or type 3 (finite-state) grammar if and only if $\overline{G}_s$ is a type 0, 1, 2, or 3 grammar. The language generated by the corresponding characteristic grammar, denoted by $L(\overline{G}_s)$, is precisely the characteristic language of $L(G_s)$. In general, a stochastic grammar G_s is said to be equivalent to some stochastic grammar G_s' if and only if $L(G_s) = L(G_s')$.

Theorem 5.1 (i) If L_s is a stochastic finite-state language, it is a stochastic context-free language. (ii) If L_s is a λ-free stochastic context-free language,† it is a stochastic context-sensitive language. (iii) If L_s is a stochastic context-sensitive language, it is a stochastic unrestricted language.

---

† A stochastic language whose characteristic language does not contain the empty string λ is called the λ-free stochastic language.

Theorem 5.2 (Chomsky normal form) Any stochastic context-free grammar is equivalent to a stochastic context-free grammar G_s in which all stochastic productions are of the forms $A \xrightarrow{p} BC$ or $A \xrightarrow{p} a$, where A, B, and C are nonterminals and a is a terminal.

Theorem 5.3 (Greibach normal form) Any stochastic context-free grammar is equivalent to a stochastic context-free grammar G_s in which all stochastic productions are of the form $A \xrightarrow{p} a\gamma$, where A is a nonterminal, a is a terminal, and γ is a string of nonterminals.

Example 5.1 Consider the stochastic finite-state grammar $G_s = (V_N, V_T, P_s, S)$, where $V_N = \{A_1, \ldots, A_k\}$, $V_T = \{a_1, \ldots, a_m\}$, $S = A_1$, and P_s, for $i = 1, 2, \ldots, k$,

$$
\begin{aligned}
&A_i \xrightarrow{p_{i1}^1} a_1 A_1 \qquad A_i \xrightarrow{p_{io}^1} a_1 \\
&A_i \xrightarrow{p_{i1}^2} a_2 A_1 \qquad \vdots \\
&\qquad \vdots \qquad\qquad A_i \xrightarrow{p_{io}^l} a_l \\
&A_i \xrightarrow{p_{ij}^l} a_l A_j \qquad \vdots \\
&\qquad \vdots \qquad\qquad A_i \xrightarrow{p_{io}^m} a_m \\
&A_i \xrightarrow{p_{ik}^m} a_m A_k
\end{aligned}
$$

Some of the probabilities p_{ij}^l, p_{io}^l may be equal to zero. Furthermore

$$
\sum_{j=1}^{k} \sum_{l=1}^{m} p_{ij}^l + \sum_{l=1}^{m} p_{io}^l = 1
$$

Specifically, let $V_N = \{S, A, B\}$, $V_T = \{0, 1\}$, and P_s:

$$
S \xrightarrow{1} 1A \qquad B \xrightarrow{0.3} 0
$$

$$
A \xrightarrow{0.8} 0B \qquad B \xrightarrow{0.7} 1S
$$

$$
A \xrightarrow{0.2} 1
$$

A typical derivation, for example, would be

$$
S \Rightarrow 1A \Rightarrow 10B \Rightarrow 100, \qquad p(100) = 1 \times 0.8 \times 0.3 = 0.24
$$

The stochastic language generated by G_s, $L(G_s)$ is as follows.

String generated x	$p(x)$
11	0.2
100	0.24
$(101)^n 11$	$0.2 \times (0.56)^n$
$(101)^n 100$	$0.24 \times (0.56)^n$

It is noted that

$$\sum_{x \in L(G_s)} p(x) = 0.2 + 0.24 + \sum_{n=1}^{\infty} (0.2 + 0.24)(0.56)^n = 1$$

5.2   Probability Measures Associated with Linear and Context-Free Grammars

Definition 5.3   If, for a stochastic grammar G_s,

$$\sum_{x \in L(G_s)} p(x) = 1 \qquad (5.4)$$

then G_s is said to be a consistent stochastic grammar.

In this section, the conditions necessary for a stochastic linear grammar and a stochastic context-free grammar to be consistent are described [2, 6, 10]. Whether or not a stochastic context-sensitive grammar is consistent under certain conditions has not yet been known. The consistency conditions for stochastic type 3 grammars are trivial, since every stochastic type 3 grammar is consistent.

5.2.1   Consistency of Stochastic Linear Grammars

Let G_s be a stochastic linear grammar with $V_N = \{A_1, \ldots, A_k\}$, $S = A_1$, and $\bar{G}_s = (V_N, V_T, P, S)$. For each $A_i \in V_N$, define the equivalence class C_{A_i} as

$$C_{A_i} = \{\text{all productions of } P \text{ with premise (left side) } A_i\}$$

Thus, the set of productions

$$P = \bigcup_{i=1}^{k} C_{A_i}$$

Let the probability associated with each production of the form $A_i \to uA_jv$, $u, v \in V_T$, be p_{ij}, and the probability associated with the production of the form $A_i \to u_r$, $u_r \in V_T$, be p_{ir}. Then

$$\sum_j p_{ij} + \sum_r p_{ir} = 1$$

The generation process can be represented as a finite-state Markov process with $(k + 1)$ states corresponding to $V_N \cup \{T\}$. The state T will correspond to an absorbing state and all the other states will correspond to transient states if G_s is consistent. A graph can be associated with each linear grammar. The nodes of the graph correspond to the elements of V_N and an additional node T corresponding to the terminal node. A path exists between the nodes A_i and A_j if and only if there is a production of the form $A_i \to uA_jv$ in P. On the other hand, there will be a path between A_i and T if and only if $A_i \to u_r$

is in P. The node corresponding to $S = A_1$ is the initial node. For every $x \in L(\bar{G}_s)$, there will be a path from the initial node through the graph to the terminal node corresponding to the derivation $S \xRightarrow{*} x$. Thus, it is noted that if G_s is consistent, then the probability of reaching the terminal node from the initial node must be 1. Otherwise, G_s would not be consistent.

To test if G_s is consistent, define the $(k + 1) \times (k + 1)$ transition matrix of the Markov process as follows:

$$M = [m_{ij}]$$

where

$$m_{ij} = \begin{cases} p_{ij}, & 1 \le i \le k, \quad 1 \le j \le (k + 1) \\ 0, & i = k + 1, \quad 1 \le j \le k \\ 1, & i = k + 1, \quad j = k + 1 \end{cases}$$

Each term m_{ij} represents the probability associated with a one-step derivation or an application of one production in the derivation of strings in $L(G_s)$. The term $m_{1, k+1}$ represents the probability of all strings in $L(G_s)$ that require the application of one production in their derivation. Let

$$M^n = \underbrace{M \times M \times \cdots \times M}_{n \text{ terms}} = [m_{ij}(n)]$$

Then, the term $m_{1, k+1}(n)$ represents the probability of all strings in $L(G_s)$ that require the application of n or fewer productions in their derivation. This leads to the following theorem.

Theorem 5.4   Let G_s be a stochastic linear grammar. G_s is consistent if and only if

$$\lim_{n \to \infty} m_{1, k+1}(n) = 1 \tag{5.5}$$

Example 5.2   Given a stochastic linear grammar $G_s = (V_N, V_T, P_s, S)$, where $V_N = \{S, A, B\}$, $V_T = \{0, 1\}$, and P_s:

$$S \xrightarrow{p_1} 0A1 \qquad A \xrightarrow{1 - p_2} 00$$

$$S \xrightarrow{1 - p_1} 1B1 \qquad B \xrightarrow{p_3} 1A1$$

$$A \xrightarrow{p_2} 0A0 \qquad B \xrightarrow{1 - p_3} 1$$

Then

$$M = \begin{array}{c} \\ S \\ A \\ B \\ T \end{array} \begin{array}{cccc} S & A & B & T \\ \left[\begin{array}{cccc} 0 & p_1 & (1 - p_1) & 0 \\ 0 & p_2 & 0 & (1 - p_2) \\ 0 & p_3 & 0 & (1 - p_3) \\ 0 & 0 & 0 & 1 \end{array}\right] \end{array}$$

Let

$$M_1 = \begin{bmatrix} 0 & 0.4 & 0.6 & 0 \\ 0 & 0.5 & 0 & 0.5 \\ 0 & 0.6 & 0 & 0.4 \\ 0 & 0 & 0 & 1 \end{bmatrix}$$

Then

$$\lim_{n \to \infty} M_1^n = \begin{bmatrix} 0 & 0 & 0 & 1 \\ 0 & 0 & 0 & 1 \\ 0 & 0 & 0 & 1 \\ 0 & 0 & 0 & 1 \end{bmatrix}$$

Thus, G_s with $p_1 = 0.4$, $p_2 = 0.5$, and $p_3 = 0.6$ is consistent, since

$$\lim_{n \to \infty} m_{1,4}(n) = 1$$

If we let

$$M_2 = \begin{bmatrix} 0 & 0.4 & 0.6 & 0 \\ 0 & 1 & 0 & 0 \\ 0 & 0.6 & 0 & 0.4 \\ 0 & 0 & 0 & 1 \end{bmatrix}$$

then

$$\lim_{n \to \infty} M_2^n = \begin{bmatrix} 0 & 0.76 & 0 & 0.24 \\ 0 & 1 & 0 & 0 \\ 0 & 0.6 & 0 & 0.4 \\ 0 & 0 & 0 & 1 \end{bmatrix}$$

Thus, G_s with $p_1 = 0.4$, $p_2 = 1$, and $p_3 = 0.6$ is not consistent, since

$$\lim_{n \to \infty} m_{1,4}(n) = 0.24 \neq 1$$

5.2.2 Consistency of Stochastic Context-Free Grammars

For a context-free grammar, all the productions are of the form

$$A \to \alpha, \qquad A \in V_N, \qquad \alpha \in V^+$$

In this case, a nonterminal at the left side of a production may directly generate zero or a finite number of nonterminals. The theory of multitype Galton–Watson branching processes [11, 12] can be applied to study the language generation process by stochastic context-free grammars. The zeroth level of a generation process corresponds to the start symbol S. The first level will be taken as β_1 where β_1 is the string generated by the production $S \to \beta_1$. The second level will correspond to the string β_2, which is obtained from β_1 by applying appropriate productions to every nonterminal in β_1. If β_1 does not contain any nonterminal, the process is terminated. Following this procedure, the jth-level string β_j is defined to be the string obtained from the string β_{j-1} by applying appropriate productions to every nonterminal of β_{j-1}. Since all nonterminals are considered simultaneously in going from the $(j-1)$th level to the jth level, only the probabilities associated with each production $A_i \rightarrow \alpha_r$, p_{ir}, need to be considered. These probabilities are handled in the same manner as was used in the stochastic linear grammar. That is, let

$$P = \bigcup_{i=1}^{k} C_{A_i}$$

For each equivalence class C_{A_i},

$$\sum_{C_{A_i}} p_{ir} = 1$$

Definition 5.4 For each C_{A_i}, $i = 1, \ldots, k$, define the k argument generating function $f_i(s_1, s_2, \ldots, s_k)$ as

$$f_i(s_1, \ldots, s_k) = \sum_{C_{A_i}} p_{ir} s_1^{\mu_{i1}(\alpha_r)} s_2^{\mu_{i2}(\alpha_r)} \cdots s_k^{\mu_{ik}(\alpha_r)} \tag{5.6}$$

where $\mu_{il}(\alpha_r)$ denotes the number of times the nonterminal A_l appears in the string α_r of the production $A_i \rightarrow \alpha_r$, and $S = A_1$.

Definition 5.5 The jth-level generating function $F_j(s_1, \ldots, s_k)$ is defined recursively as

$$F_0(s_1, \ldots, s_k) = s_1$$
$$F_1(s_1, \ldots, s_k) = f_1(s_1, \ldots, s_k)$$
$$F_j(s_1, \ldots, s_k) = F_{j-1}(f_1(s_1, \ldots, s_k), \ldots, f_k(s_1, \ldots, s_k)) \tag{5.7}$$

Example 5.3 A grammar $G_s = (V_N, V_T, P_s, S)$, where $V_N = \{A_1, A_2\}$, $V_T = \{a, b\}$, $S = A_1$, and P_s:

$$A_1 \xrightarrow{p_{11}} aA_1A_2 \qquad A_2 \xrightarrow{p_{21}} aA_2A_2$$

$$A_1 \xrightarrow{p_{12}} b \qquad A_2 \xrightarrow{p_{22}} aa$$

Then the generation process for the string *aabaaaa* in terms of levels would be

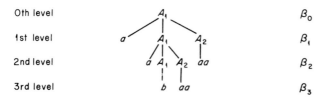

The generating functions for C_{A_1} and C_{A_2} are, respectively,

$$f_1(s_1, s_2) = p_{11}s_1s_2 + p_{12}, \qquad f_2(s, s_2) = p_{21}s_2{}^2 + p_{22}$$

The jth-level generating functions for $j = 0, 1, 2$ are

$$F_0(s_1, s_2) = s_1$$
$$F_1(s_1, s_2) = f_1(s_1, s_2) = p_{11}s_1s_2 + p_{12}$$
$$F_2(s_1, s_2) = F_1(f_1(s_1, s_2), f_2(s_1, s_2))$$
$$= p_{11}f_1(s_1, s_2)f_2(s_1, s_2) + p_{12}$$
$$= p_{11}^2p_{21}s_1s_2{}^3 + p_{11}^2p_{22}\,s_1s_2 + p_{11}p_{12}\,p_{21}s_2{}^2 + p_{11}p_{12}\,p_{22} + p_{12}$$

After examining the previous example, we can express $F_j(s_1, \ldots, s_k)$ as

$$F_j(s_1, \ldots, s_k) = G_j(s_1, \ldots, s_k) + K_j$$

where $G_j(s_1, \ldots, s_k)$ denotes the polynomial of $s_1, \ldots, s_k$ without including the constant term. The constant term K_j corresponds to the probability of all the strings $x \in L(\bar{G}_s)$ that can be derived in j or fewer levels. This leads to the following theorem.

Theorem 5.5   A stochastic context-free grammar G_s is consistent if and only if

$$\lim_{j \to \infty} K_j = 1 \tag{5.8}$$

Note that if the foregoing limit does not equal 1, there is a finite probability that a generation process may never terminate. Thus the probability measure defined over L will be less than 1 and, consequently, G_s will not be consistent. On the other hand, if the limit is equal to 1, then there exists no such infinite (nonterminating) generation process, since the limit represents the probability of all the strings that are generated by applications of a finite number of productions. Consequently, G_s is consistent. The problem of testing the consistency of a given stochastic context-free grammar can be solved by using the following testing procedure developed for branching processes.

Definition 5.6   The expected number of occurrences of the nonterminal A_j in the production set C_{A_i} is

$$e_{ij} = \left. \frac{\partial f_i(s_1, \ldots, s_k)}{\partial s_j} \right|_{s_1, \ldots, s_k = 1} \tag{5.9}$$

Definition 5.7   The first moment matrix E of the generation process corresponding to a stochastic context-free grammar G_s is defined as

$$E = [e_{ij}], \qquad 1 \le i, j \le k \tag{5.10}$$

where k is the number of nonterminals in G_s.

Theorem 5.6   For a given stochastic context-free grammar G_s, order the eigenvalues or the characteristic roots $\rho_1, \ldots, \rho_k$ of the first moment matrix according to the descending order of their magnitudes; that is,

$$|\rho_i| \geq |\rho_j| \quad \text{if} \quad i < j \tag{5.11}$$

Then G_s is consistent if $\rho_1 < 1$ and is not consistent if $\rho_1 > 1$.

Example 5.4   For the stochastic context-free grammar G_s given in Example 5.3,

$$e_{11} = \frac{\partial f_1(s_1, s_2)}{\partial s_1}\bigg|_{s_1, s_2 = 1} = p_{11}$$

$$e_{12} = \frac{\partial f_1(s_1, s_2)}{\partial s_2}\bigg|_{s_1, s_2 = 1} = p_{11}$$

$$e_{21} = \frac{\partial f_2(s_1, s_2)}{\partial s_1}\bigg|_{s_1, s_2 = 1} = 0$$

$$e_{22} = \frac{\partial f_2(s_1, s_2)}{\partial s_2}\bigg|_{s_1, s_2 = 1} = 2p_{21}$$

Thus,

$$E = \begin{bmatrix} p_{11} & p_{11} \\ 0 & 2p_{21} \end{bmatrix}$$

The characteristic equation associated with E is

$$\Phi(x) = (x - p_{11})(x - 2p_{21})$$

Hence, G_s will be consistent as long as $p_{21} < \frac{1}{2}$.

5.3   Languages Accepted by Stochastic Automata

Based on the concept of stochastic finite-state automata, Rabin [13], Paz [14], Turakainen [15], and Salomaa [16] have studied a class of Λ-stochastic finite-state languages (or simply Λ-stochastic languages), and Fu and Li [17] have defined a class of maximum-likelihood stochastic languages. Parallel to the formulation of stochastic finite-state automata and Λ-stochastic languages, Ellis [18], and Huang and Fu [19] have defined a class of Λ-stochastic pushdown languages, which are the languages accepted by stochastic pushdown automata with a cutpoint Λ, such that $0 \leq \Lambda < 1$. In this section we shall give the basic formulation of stochastic automata (or acceptors) and define the class of languages accepted by stochastic automata.

5.3.1   Languages Accepted by Stochastic Finite-State Automata

Definition 5.8   A stochastic finite-state automaton (or simply, a stochastic automaton) is a five-tuple $S_a = (\Sigma, Q, M, \pi_0, F)$ where Σ and Q are finite sets of input symbols and internal states, respectively; M is a mapping of Σ into the set of $n \times n$ (where n is the number of states in Q) stochastic state transition matrices; π_0 is an n-dimensional row vector and is designated as the initial state distribution; $F \subseteq Q$ is a finite set of final states.

The interpretation of $M(a)$, $a \in \Sigma$, can be stated as follows. Let $M(a) = [p_{ij}(a)]$ where $p_{ij}(a) \geq 0$ is the probability of entering state q_j from state q_i under the input a, and $\sum_{j=1}^{n} p_{ij} = 1$, for all $i = 1, \ldots, n$. The domain of M can be extended from Σ to Σ^* by defining

(1) $M(\lambda) = I$ where λ is the empty string and I is an $n \times n$ identity matrix;
(2) $M(a_1, \ldots, a_k) = M(a_1), \ldots, M(a_k)$ where $k > 1$ and all $a_i \in \Sigma$.

The weighted language accepted by a stochastic finite-state automaton $S_a = (\Sigma, Q, M, \pi_0, F)$ is $T(S_a) = \{(x, p(x)) \mid x \in \Sigma^* \text{ and } p(x) = \pi_0 M(x)\pi_F > 0\}$. The language accepted by S_a with a cutpoint Λ, such that $0 \leq \Lambda < 1$, is

$$L(S_a, \Lambda) = \{x \mid x \in \Sigma^* \text{ and } \pi_0 M(x)\pi_F > \Lambda\} \qquad (5.12)$$

where π_F is an n-dimensional column vector where the ith component is equal to 1 if $q_i \in F$ and 0 otherwise. $L(S_a, \Lambda)$ is called the Λ-stochastic finite-state language or simply the Λ-stochastic language.

Examples 5.5   Consider a stochastic finite-state automaton $S_a = (\Sigma, Q, M, \pi_0, F)$ where $\Sigma = \{0, 1\}$, $Q = \{q_1, \ldots, q_6\}$,

$$M(0) = \begin{bmatrix} 0 & 0 & 0 & 0 & 0 & 1 \\ 0 & 0 & 0 & \frac{1}{2} & \frac{1}{2} & 0 \\ 0 & 0 & 0 & \frac{1}{2} & 0 & \frac{1}{2} \\ 0 & 0 & 0 & 0 & 0 & 1 \\ 0 & 0 & 0 & 0 & 0 & 1 \\ 0 & 0 & 0 & 0 & 0 & 1 \end{bmatrix}, \quad M(1) = \begin{bmatrix} 0 & 1 & 0 & 0 & 0 & 0 \\ 0 & \frac{1}{2} & \frac{1}{2} & 0 & 0 & 0 \\ 0 & 0 & 1 & 0 & 0 & 0 \\ 0 & 0 & 0 & \frac{1}{2} & \frac{1}{2} & 0 \\ 0 & 0 & 0 & 0 & 1 & 0 \\ 0 & 0 & 0 & 0 & 0 & 1 \end{bmatrix}$$

$$\pi_0 = [1 \quad 0 \quad 0 \quad 0 \quad 0 \quad 0], \qquad F = \{q_5\}$$

Then

$$\pi_F = [0 \quad 0 \quad 0 \quad 0 \quad 1 \quad 0]^T$$

and

$$p(x) = \pi_0 M(x)\pi_F = \begin{cases} \frac{1}{2} + (\frac{1}{2})^m - (\frac{1}{2})^{n+1}, & x = 1^m 0 1^n, \quad m, n > 0 \\ 0, & \text{otherwise} \end{cases}$$

Hence,

$$T(S_a) = \{(1^m 01^n, p(x)) \mid m, n > 0\}$$

and

$$L(S_a, \tfrac{1}{2}) = \{1^m 01^n \mid 0 < m < n\}$$

which is a context-free language.

For a given stochastic finite-state grammar G_s, it is possible to construct a stochastic finite-state automaton S_a accepting $L(G_s)$ [17]. In the following, the procedure for constructing the stochastic finite automaton S_a from a given stochastic grammar G_s is described.

Let $G_s = (V_N, V_T, P_s, S)$ and $S_a = (\Sigma, Q, M, \pi_0, F)$.

(1) The input set Σ of S_a is equal to the set of terminal variables V_T; that is, $\Sigma = V_T$.

(2) The state set Q of S_a is equal to the union of the set of nonterminals V_N and two added states T, R, corresponding to the states of termination and rejection, respectively. The rejection state is mainly for normalization purposes. Let $T = A_{k+1}$ and $R = A_{k+2}$, then $Q = V_N \cup \{T, R\}$.

(3) π_0 is a row vector with its components equal to 1 in the position of the S state, others equal to 0.

(4) The set of final states has only one element, that is, T, the termination state.

(5) The state transition matrices M are formed on the basis of the stochastic productions P_s of the grammar. (Refer to Example 5.1.) If in P_s there is a production $A_i \rightarrow a_l A_j$ with probability p^l_{ij}, then for state transition matrix $M(a_l)$, the entry in the ith row and the jth column is p^l_{ij}. For a production $A_i \rightarrow a_l$ with probability p^l_{i0}, the entry in the ith row and the $(k+1)$th column in $M(a_l)$ is p^l_{i0}. The entries in the $(k+2)$th column are so determined that the state transition matrix is always stochastic.

Example 5.6   Consider the stochastic finite-state grammar G_s given in Example 5.1. Clearly, according to the procedure described above, we have $\Sigma = \{1, 0\}$, $Q = \{S, A, B, T, R\}$, $\pi_0 = [1 \quad 0 \quad 0 \quad 0 \quad 0]$, and $F = \{T\}$. Moreover,

$$M(1) = \begin{array}{c} \\ S \\ A \\ B \\ T \\ R \end{array}\begin{array}{ccccc} S & A & B & T & R \\ \left[\begin{array}{ccccc} 0 & 1 & 0 & 0 & 0 \\ 0 & 0 & 0 & 0.2 & 0.8 \\ 0.7 & 0 & 0 & 0 & 0.3 \\ 0 & 0 & 0 & 0 & 1 \\ 0 & 0 & 0 & 0 & 1 \end{array}\right] \end{array}, \qquad M(0) = \begin{array}{c} \\ S \\ A \\ B \\ T \\ R \end{array}\begin{array}{ccccc} S & A & B & T & R \\ \left[\begin{array}{ccccc} 0 & 0 & 0 & 0 & 1 \\ 0 & 0 & 0.8 & 0 & 0.2 \\ 0 & 0 & 0 & 0.3 & 0.7 \\ 0 & 0 & 0 & 0 & 1 \\ 0 & 0 & 0 & 0 & 1 \end{array}\right] \end{array}$$

The probability that a string x is generated is equal to $\pi_0 M(x)\pi_F$. In this example, the probability that the string 10111 is generated is equal to

$$\pi_0 M(1) \cdot M(0)M^3(1)\pi_F = 0.112$$

Therefore, the stochastic finite-state automaton is also a convenient tool for computing the probability associated with each string in the language generated by the corresponding stochastic finite-state grammar.

5.3.2 Maximum-Likelihood Stochastic Languages

Based on the concept of the maximum likelihood of the final state distribution, the class of maximum-likelihood stochastic languages was formally introduced by Fu and Li [17]. Its formal definition is defined as follows.

Definition 5.9 For a given stochastic finite-state automaton $S_a = (\Sigma, Q, M, \pi_0, F)$, let

$$\pi_0 M(x) = (p(q_1|x), \ldots, p(q_n|x))$$

where n is the number of states in Q. Let

$$g(x) = \max_i \{p(q_i|x)\} \qquad \text{and} \qquad \bar{Q} = \{q | p(q|x) = g(x)\}$$

where $g(x)$ is called the maximum-likelihood membership function (MLMF) of $x \in \Sigma^*$. The class of maximum-likelihood stochastic languages (mlsl) accepted by the stochastic finite-state automaton S_a is defined to be

$$L_m(S_a) = \{x | \bar{Q} \cap F \neq \varnothing\} \tag{5.13}$$

where $\varnothing$ is the empty set.

Example 5.7 Consider a stochastic finite-state automaton $S_a = (\Sigma, Q, M, \pi_0, F)$, where $\Sigma = \{0, 1\}$, $Q = \{q_1, q_2, q_3, q_4\}$,

$$M(0) = \begin{bmatrix} \frac{1}{2} & \frac{1}{2} & 0 & 0 \\ 0 & 1 & 0 & 0 \\ 0 & 0 & 1 & 0 \\ 0 & 0 & 0 & 1 \end{bmatrix}, \qquad M(1) = \begin{bmatrix} 1 & 0 & 0 & 0 \\ 0 & 1 & 0 & 0 \\ 0 & 0 & \frac{1}{2} & \frac{1}{2} \\ 0 & 0 & 0 & 1 \end{bmatrix}$$

$$\pi_0 = [\tfrac{1}{2} \quad 0 \quad \tfrac{1}{2} \quad 0], \qquad F = \{q_4\}$$

Let the sequence $x \in \Sigma^*$ such that the number of 0's and the number of 1's are denoted by k_1 and k_2, respectively. It can be shown that

$$\pi_0 M(x) = [(\tfrac{1}{2})^{k_1+1}, \tfrac{1}{2} - (\tfrac{1}{2})^{k_1+1}, (\tfrac{1}{2})^{k_2+1}, \tfrac{1}{2} - (\tfrac{1}{2})^{k_2+1}]$$

and

$$L_m(S_a) = \{x \in \Sigma^* | k_2 > k_1 > 0\}$$

which is a context-free language.

5.3.3   Languages Accepted by Stochastic Pushdown Automata

Definition 5.10   A *stochastic pushdown automaton* (spda) is a seven-tuple $M_s = (\Sigma, Q, \Gamma, \delta_s, q_0, Z_0, F)$ in which

(1) Σ is a finite set of input symbols;
(2) Q is a finite set of states;
(3) Γ is a finite set of pushdown symbols;
(4) $f_0 \in Q$ is the initial state;
(5) $Z_0 \in \Gamma$ is the symbol initially appearing on the top of the pushdown storage;
(6) $F \subseteq Q$ is the set of final states;
(7) $\delta_s: Q \times (\Sigma \cup \{\lambda\}) \times \Gamma \xrightarrow{\;p\;} 2^{Q \times \Gamma^*}$ where p is a probability measure defined over $2^{Q \times \Gamma^*}$ for each element in $Q \times (\Sigma \cup \{\lambda\}) \times \Gamma$.

δ_s consists of two types of stochastic mappings:

$$(i) \quad \delta_s(q, a, Z) = \{(q_i, \gamma_i, p_i), i = 1, \ldots, m\} \tag{5.14}$$

where

$$\sum_{i=1}^{m} p_i = 1, \qquad p_i > 0, \quad i = 1, \ldots, m$$

This type of mapping can be interpreted as follows. If M_s is currently in state q with a being scanned by the input head and Z appearing on the top of the pushdown storage, M_s can enter state q_i and replace Z by γ_i, all with probability p_i. The input head is advanced to the next symbol.

$$(ii) \quad \delta_s(q, \lambda, Z) = \{(q_i, \gamma_i, p_i), \quad i = 1, \ldots, m\} \tag{5.15}$$

where

$$\sum_{i=1}^{m} p_i = 1 \quad \text{and} \quad p_i > 0, \qquad i = 1, \ldots, m$$

The interpretation of this type of stochastic mapping can be stated as follows. If M_s is currently in state q with Z appearing at the top of the pushdown storage, independently of the symbol being scanned by the input head, it can enter state q_i and replace Z by γ_i all with probability p_i. The input head is not advanced. This type of stochastic mapping allows M_s to manipulate its pushdown storage without scanning any input.

Definition 5.11   A configuration of a spda M_s is defined as a triple (q, γ, p) where $q \in Q$, $\gamma \in \Gamma^*$, and p is defined as the probability of entering state q with γ on the pushdown storage by means of a mapping defined in δ_s from an initial configuration $(q_0, Z_0, 1)$.

If $\delta_s(q, a, Z) = \{(q_i, \gamma_i, p_i), i = 1, \ldots, m\}$ is a mapping defined in M_s, then it can be equivalently expressed as

$$a: (q, Z\beta, p) \xrightarrow[M_s]{} (q_i, \gamma_i \beta, pp_i), \qquad i = 1, \ldots, m$$

The interpretation is that with probability p_i spda M_s enters configuration $(q_i, \gamma_i \beta, pp_i)$ from configuration $(q, Z\beta, p)$ under the input a. This definition of the relation $\xrightarrow[M_s]{}$ can be generalized as follows: If

$$a_i: \quad (q_{i-1}, \gamma_{i-1}, p_{i-1}) \xrightarrow[M_s]{} (q_i, \gamma_i, p_{i-1}p_i)$$

is defined for $i = 1, \ldots, n$, then it can be equivalently expressed as

$$a_1 \cdots a_n: \quad (q_0, \gamma_0, p_0) \xrightarrow[M_s]{*} \left(q_n, \gamma_n, \prod_{i=0}^{n} p_i\right)$$

where $\prod_{i=1}^{n} p_i$ is the probability of entering configuration $(q_n, \gamma_n, \prod_{i=0}^{n} p_i)$ from configuration (q_0, γ_0, p_0) under the input sequence $a_1 \cdots a_n$ by means of a derivation defined in δ_s.

The weighted pushdown language accepted by spda $M_s = (\Sigma, Q, \Gamma, \delta_s, q_0, Z_0, F)$ can be defined equivalently as follows.

(1)   By final state, the weighted pushdown language is

$$T(M_s) = \left\{(x, p(x)) \mid x \in \Sigma^*: (q_0, Z_0, 1) \xrightarrow[M_s]{*} (q_i, \gamma_i, p_i)\right.$$

$$\text{for} \quad \gamma_i \in \Gamma^*, \quad q_i \in F, \quad i = 1, \ldots, k, \qquad (5.16)$$

$$\left. \text{and} \quad p(x) = \sum_{i=1}^{k} p_i\right\}$$

where k is the number of distinctively different transitions defined in M_s.

(2)   By empty store, the weighted pushdown language is

$$N(M_s) = \left\{(x, p(x)) \mid x \in \Sigma^*: (q_0, Z_0, 1) \xrightarrow[M_s]{*} (q_i, \lambda, p_i)\right.$$

$$\text{for} \quad q_i \in Q, \quad i = 1, \ldots, k, \qquad (5.17)$$

$$\left. \text{and} \quad p(x) = \sum_{i=1}^{k} p_i\right\}$$

where k is the number of distinctively different transitions defined in M_s.

The language accepted by spda $M_s = (\Sigma, Q, \Gamma, \delta_s, q_0, Z_0, F)$ with a cutpoint Λ, such that $0 \le \Lambda < 1$ can be defined as follows:

(1)   By final state, the language is

$$T(M_s \Lambda) = \left\{ x \in \Sigma^* \mid x \colon (q_0, Z_0, 1) \xrightarrow[M_s]{*} (q_i, \gamma_i, p_i) \right.$$

$$\text{for} \quad \gamma \in \Gamma^*, \quad q_i \in F, \quad i = 1, \ldots, k, \qquad (5.18)$$

$$\left. \text{and} \quad \sum_{i=1}^{k} p_i = p(x) > \Lambda \right\}$$

where k is the number of all distinctively different transitions.

(2)   By empty store, the language is

$$N(M_s, \Lambda) = \left\{ x \in \Sigma^* \mid x \colon (q_0, Z_0, 1) \xrightarrow[M_s]{*} (q_i, \lambda, p_i) \right.$$

$$\text{for} \quad q_i \in Q, \quad i = 1, \ldots, k, \qquad (5.19)$$

$$\left. \text{and} \quad \sum_{i=1}^{k} p_i = p(x) > \Lambda \right\}$$

where k is the number of all distinctively different transitions.

It can be shown that $T(M_s, \Lambda)$ and $N(M_s, \Lambda)$ are equivalent and can be called the Λ-stochastic pushdown languages. The properties of Λ-stochastic pushdown languages will be discussed in Section 5.3.5.

Example 5.8   Let $M_s = (\{a, b\}, Q, \{A, Z_0\}, \delta, q_0, Z_0, \{q_4, q_{11}\})$ be a spda where

$$Q = \{q_0, q_1, q_2, q_3, q_4, q_5, q_6, q_7, q_8, q_9, q_{10}, q_{11}\}$$

and δ is defined as follows:

$$\delta(q_0, a, Z_0) = \{(q_1, AZ_0, \tfrac{1}{2}), (q_6, AZ_0, \tfrac{1}{2})\}$$

$\delta(q_1, a, A) = \{(q_1, AA, 1)\}$	$\delta(q_6, b, A) = \{(q_7, A, 1)\}$
$\delta(q_1, b, A) = \{(q_2, \lambda, 1)\}$	$\delta(q_7, b, A) = \{(q_7, A, 1)\}$
$\delta(q_2, b, A) = \{(q_2, \lambda, \tfrac{1}{2}), (q_3, \lambda, \tfrac{1}{2})\}$	$\delta(q_7, a, A) = \{(q_8, \lambda, 1)\}$
$\delta(q_3, b, A) = \{(q_3, \lambda, 1)\}$	$\delta(q_8, a, A) = \{(q_8, \lambda, \tfrac{1}{2}), (q_9, \lambda, \tfrac{1}{2})\}$
$\delta(q_2, a, Z_0) = \{(q_4, Z_0, 1)\}$	$\delta(q_9, a, A) = \{(q_9, \lambda, 1)\}$
$\delta(q_3, a, Z_0) = \{(q_5, Z_0, 1)\}$	$\delta(q_8, a, Z_0) = \{(q_{10}, Z_0, 1)\}$
$\delta(q_4, a, Z_0) = \{(q_4, Z_0, 1)\}$	$\delta(q_9, a, Z_0) = \{(q_{11}, Z_0, 1)\}$
$\delta(q_5, a, Z_0) = \{(q_6, Z_0, 1)\}$	$\delta(q_{11}, a, Z_0) = \{(q_{11}, Z_0, 1)\}$
$\delta(q_6, a, A) = \{(q_6, AA, 1)\}$	$\delta(q_{10}, a, Z_0) = \{(q_{10}, Z_0, 1)\}$

Then $T(M_s) = \{(a^k b^m a^n, p(a^k b^m a^n)) \mid k, m, n > 0\}$, where

$$p(a^k b^m a^n) = \begin{cases} (\tfrac{1}{2})^k + \tfrac{1}{2} - (\tfrac{1}{2})^{k+1}, & k = m, \quad n > k \\ (\tfrac{1}{2})^k, & k = m, \quad n \le k \\ \tfrac{1}{2} - (\tfrac{1}{2})^{k+1}, & k \ne m, \quad n > k \\ 0, & k \ne m, \quad n \le k \end{cases}$$

and

$$T(M_s, \tfrac{1}{2}) = \{a^k b^k a^m \mid m > k \ge 1\}$$

which is not a context-free language.

5.3.4 Properties of Languages Accepted by Stochastic Finite-State Automata

The relationships between finite-state languages and finite-state automata have been shown in Chapter 2. Some basic relationships between stochastic finite-state languages, weighted finite-state languages, and the weighted languages accepted by stochastic finite-state automata are summarized as follows.

(F1) Let G_s be a stochastic finite-state grammar. Then there exists a stochastic finite-state automaton S_a such that $L(G_s) = T(S_a)$.

(F2) The class of weighted languages accepted by stochastic finite-state automata properly contains the class of stochastic finite-state languages and is properly contained within the class of weighted finite-state languages.

The properties of Λ-stochastic (finite-state) languages can be summarized as follows.

(S1) Every Λ-stochastic language is a Λ_1-stochastic language, for any Λ_1, such that $0 < \Lambda_1 > 1$. In other words, every Λ-stochastic language is $L(S_a, \tfrac{1}{2})$, for some stochastic finite automaton S_a.

(S2) The union $L_1 \cup L_2$, the intersection $L_1 \cap L_2$, and the difference $L_1 - L_2$ (or $L_1 \cap \bar{L}_2$) of a Λ-stochastic language L_1 and a finite-state language L_2 are all Λ-stochastic languages, where $\bar{L}_2$ is the complement of L_2.

(S3) There exists a stochastic finite-state automaton and $0 < \Lambda < 1$ such that $L(S_a, \Lambda)$ is not a finite-state language, that is, it is not accepted by any deterministic finite-state automaton.

(S4) Let $\{Q \mid Qx \in L\}$ and $\{Q \mid xQ \in L\}$ be the right and left derivatives of a language L with respect to a string x. Then the right and left derivatives of Λ-stochastic languages are Λ-stochastic languages.

(S5) Let the cutpoint Λ be isolated with respect to a stochastic finite-state automaton S_a; that is, there exists a $\delta > 0$ such that $|\pi_0 M(x)\pi_F - \Lambda| \ge \delta$

for all $x \in \Sigma^*$. Then there exists a deterministic finite-state automaton which will accept $L(S_a, \Lambda)$. That is, $L(S_a, \Lambda)$ is a finite-state language.

(S6) There exists a context-free language which is not a Λ-stochastic language.

The properties of maximum-likelihood stochastic languages (mlsl) can be summarized as follows.

(M1) The class of mlsl properly includes the class of finite-state languages.

(M2) Let S_a be a stochastic finite-state automaton with n states. Then $L(S_a, \Lambda)$ with $\Lambda \leq 1/n$ properly contains mlsl.

(M3) All left derivations of a mlsl are mlsl.

(M4) If $g_1, \ldots, g_k$ are MLMF, so is $(g_1 \cup \cdots \cup g_k)/k$.

(M5) Let S_a be a two-state stochastic automaton. Then $L_m(S_a) = L(S_a, \frac{1}{2})$.

(M6) The intersection of the class of context-free languages, Λ-stochastic languages, and mlsl are not empty.

5.3.5 Properties of Languages Accepted by Stochastic Pushdown Automata

The relationships between pushdown automata and context-free languages have been shown in Chapter 2. Some basic results on the properties of weighted pushdown languages are summarized as follows.

(W1) A weighted pushdown language L_{wp} is $N(M_s)$ for some spda M_s if and only if L_{wp} is $T(M_s')$ for some spda M_s'.

(W2) Let G_s be a stochastic context-free grammar. Then there exists a spda M_s such that $L(G_s) = N(M_s)$.

(W1) implies that the weighted pushdown language can be equivalently defined by either final state or empty store, while (W2) implies that the set of weighted pushdown languages contains the class of stochastic context-free languages. It is easy to show that the set of weighted pushdown languages is properly contained within the class of weighted context-free languages.

The properties of the class of languages accepted by a spda with a cutpoint Λ, such that $0 \leq \Lambda < 1$, can be summarized as follows.

(P1) A language L is $N(M_s, \Lambda)$ for some spda M_s with a cutpoint Λ, such that $0 \leq \Lambda < 1$, if and only if L is $T(M_s', \Lambda)$ for some spda M_s'.

(P2) Let M_s be a spda. Then $N(M_s, 0)$ defines a context-free language.

(P3) Let M_s be a spda. The class of languages accepted by M_s with a cutpoint Λ, such that $0 \leq \Lambda < 1$, properly contains the class of context-free languages.

(P4) Every Λ-stochastic pushdown language is a Λ'-stochastic pushdown language, for any Λ' such that $0 < \Lambda' < 1$.

(P5)   The union $L_1 \cup L_2$, the intersection $L_1 \cap L_2$, and the difference $L_1 - L_2$ (or $L_1 \cap \bar{L}_2$) of a Λ-stochastic pushdown language L_1 and a deterministic language L_2 are Λ-stochastic pushdown languages [20, 21].

5.4   Stochastic Programmed and Indexed Grammars

Both programmed grammars [22] and indexed grammars [23] are convenient methods by which to generate a broad class of languages. We shall extend the formulations of programmed grammars and indexed grammars to stochastic programmed and indexed grammars.

5.4.1   Stochastic Programmed Grammars and Languages†

Definition 5.12   A stochastic programmed grammar (spg) is a five-tuple $G_{sp} = (V_N, V_T, J, P_s, S)$ where

(1) V_N is a finite set of nonterminals;
(2) V_T is a finite set of terminals;
(3) J is a finite set of production labels;
(4) P_s is a finite set of productions; and
(5) S is the start symbol.

Each production in P_s is of the form

$$(r)   \omega \to \eta   S(U)P(U)   F(W)P(W)$$

where $\omega \to \eta$ is the core of the production for $\omega \in (V_N \cup V_T)^*V_N(V_N \cup V_T)^*$ and $\eta \in (V_N \cup V_T)^*$.

Each production in P_s has a distinct label, $r \in J$. $U \subseteq J$ and $W \subseteq J$ are called the success and the failure go-to fields, respectively. $P(U)$ and $P(W)$ are probability distributions associated with the sets U and W, respectively. In applying the production to an intermediate string ξ derived from S under the leftmost interpretation, if ξ contains the substring ω, then the leftmost occurrence of ω is expanded into η and the next production is selected from the success go-to field U. If ξ does not contain ω, the ξ is not changed and the next production is selected from the failure go-to field W.
Suppose that

$$S = \omega_1 \overset{r_1}{\Longrightarrow} \omega_2 \overset{r_2}{\Longrightarrow} \cdots \overset{r_n}{\Longrightarrow} \omega_{n+1} = x$$

is a derivation of x from S in which r_i is the label of the production used to directly generate ω_{i+1} from ω_i. The probability p associated with this derivation of x from S is defined as the product of the conditional probabilities

$$p(x) = p(r_1)p(r_2|r_1) \cdots p(r_n|r_{n-1})$$

† See Souza [7] and Swain and Fu [8].

The interpretation of the conditional probability $p(r_i|r_{i-1})$ is the probability associated with the selection of production r_i as the next production from the success go-to field if production r_{i-1} is successfully applied. Otherwise, $p(r_i|r_{i-1})$ denotes the probability associated with the selection of production r_i as the next production from the failure go-to field of the r_{i-1}th production.

Definition 5.13   A stochastic unconditional transfer programmed grammar (sutpg) is a stochastic programmed grammar in which $U = W$ and $P(U) = P(W)$ for each production defined in the grammar.

Definition 5.14   Let G_{sp} be a stochastic programmed grammar. The corresponding characteristic programmed grammar, denoted by $\bar{G}_{sp}$, is formed from G_{sp} by deleting the probability distribution associated with each go-to field. The language generated by $\bar{G}_{sp}$ is called the programmed language [22].

Definition 5.15   A stochastic programmed grammar G_{sp} is said to be a stochastic context-sensitive, context-free, or finite-state programmed grammar if $\bar{G}_{sp}$ is a context-sensitive, context-free, or finite-state programmed grammar.

Definition 5.16   A stochastic context-free programmed grammar (scfpg) operating under leftmost derivation is one for which each production is applied to the leftmost nonterminal of the current intermediate string.

Definition 5.17   A stochastic context-free programmed grammar operating under leftmost interpretation is one for which each production is applied to the leftmost occurrence of the nonterminal which it expands.

The distinction between leftmost derivation and leftmost interpretation can be clearly illustrated by the following example. Suppose that the current intermediate string is $\omega = aABAB$, where a is a terminal and A and B are nonterminals, and the next production to be applied to ω is $B \rightarrow b$. This production will succeed in expanding the leftmost B in ω into b under leftmost interpretation, while it will fail in expanding B in ω into b under leftmost derivation because B is not the leftmost nonterminal.

Definition 5.18   A stochastic context-free programmed grammar operating under rightmost interpretation is one for which each production is applied to the rightmost occurrence of the nonterminal which it expands.

The stochastic programmed language generated by a stochastic programmed grammar $G_{sp} = (V_N, V_T, J, P_s, S)$ is defined as

$$L(G_{sp}) = \left\{ (x, p(x)) \,|\, x \in V_T^*, \; S \overset{p_i}{\underset{*}{\Longrightarrow}} x, \qquad \text{for} \quad i = 1, \ldots, k, \right.$$

$$\left. \text{and} \quad p(x) = \sum_{i=1}^{k} p_i \right\} \qquad (5.20)$$

where k is the number of all distinctively different derivations of x from S, and p_i is the probability associated with the ith distinctive derivation of x. A stochastic programmed grammar G_{sp} is said to be consistent if and only if

$$\sum_{x \in L(G_{sp})} p(x) = 1$$

Example 5.9 Consider the stochastic context-free programmed grammar $G_{sp} = (V_N, V_T, J, P_s, S)$, where $V_N = \{S, A, B, C, D\}$, $V_T = \{a, b, c, d\}$, $J = \{1, 2, 3, 4, 5, 6, 7\}$, and P_s:

Label	Core	U	P(U)	W	P(W)
1	$S \rightarrow aAB$	2, 3	$\alpha, 1 - \alpha$	$\varnothing$	1
2	$A \rightarrow aAC$	2, 3	$1 - \beta, \beta$	$\varnothing$	1
3	$A \rightarrow D$	4	1	$\varnothing$	1
4	$C \rightarrow d$	5	1	6	1
5	$D \rightarrow bDc$	4	1	$\varnothing$	1
6	$B \rightarrow d$	7	1	$\varnothing$	1
7	$D \rightarrow bc$	$\varnothing$	1	$\varnothing$	1

A typical derivation will be

$$S \xoverset{1}{\Longrightarrow} aAB \xoverset{2}{\Longrightarrow} aaACB \xoverset{3}{\Longrightarrow} aaDCB$$

$$\xoverset{4}{\Longrightarrow} aaDdB \xoverset{5}{\Longrightarrow} aabDcdB \xoverset{6}{\Longrightarrow} aabDcdd \xoverset{7}{\Longrightarrow} aabbccdd$$

and $p(aabbccdd) = \alpha\beta$. In general, $p(a^n b^n c^n d^n) = \alpha\beta(1 - \beta)^{n-2}, n = 2, 3 \ldots$. The properties of stochastic programmed grammars and languages can be summarized as follows.

Theorem 5.7 The set of stochastic programmed languages generated by stochastic context-free programmed grammars (scfpg) operating under the rightmost interpretation is identical to the set of stochastic programmed languages generated by scfpg operating under the leftmost interpretation.

Theorem 5.8 Any scfpg operating under the leftmost derivation can be put into an equivalent form with no failure go-to field and a single nonterminal look-ahead.

Theorem 5.9 The set of stochastic programmed languages generated by a scfpg operating under leftmost derivation is precisely the set of stochastic context-free languages.

Theorem 5.10 The set of stochastic programmed languages generated by a scfpg operating under leftmost interpretation properly contains the set of stochastic context-free languages.

5.4.2   *Stochastic Indexed Grammars and Languages*†

Indexed grammars provide a means of generating a class of languages which properly includes the class of context-free languages and is properly included within the class of context-sensitive languages. It has been shown [23] that the class of indexed languages is precisely the class of languages accepted by one-way nondeterministic nested stack automata. In this section, stochastic indexed grammars, stochastic indexed languages, and stochastic nested stack automata are presented.

Definition 5.19   A stochastic indexed grammar is a five-tuple $G_{si} = (V_N, V_T, F, P_s, S)$ where

(1) V_N is a set of nonterminals
(2) V_T is a set of terminals
(3) F is a finite set of elements each of which is a finite set of stochastic productions of the form $A \xrightarrow{p} \omega$, where $A \in V_N$, $\omega \in (V_N \cup V_T)^*$, and p is the probability of applying the production $A \to \omega$ to expand A into ω. An element $f \in F$ is called an index (or flag), and $A \xrightarrow{p} \omega$ in f is called a stochastic indexed production. Let $\{A \xrightarrow{p_i} \omega_i, i = 1, \ldots, l\}$ be the set of all stochastic indexed productions with A on the left-hand side defined in $f \in F$. Then

$$\sum_{i=1}^{l} p_i = 1$$

(4) P_s is a finite set of stochastic productions of the form $A \xrightarrow{p} \gamma$ where $A \in V_N$, $\gamma \in (V_N F^* \cup V_T)^*$, and p is the probability of applying $A \longrightarrow \gamma$ to expand A into γ. Let the set of all stochastic productions in P_s with A on the left be

$$\{A \xrightarrow{p_i} \gamma_i, i = 1, \ldots, k\}$$

Then

$$\sum_{i=1}^{k} p_i = 1$$

(5) $S \in V_N$ is the start symbol.

The use of the stochastic production P_s and stochastic indexed production F in a derivation can be described as follows. Let $A \in V_N$, $X_i \in (V_N \cup V_T)$, and $\psi_i \in F^*$ for $i = 1, \ldots, k$. (i) Suppose that $A \xrightarrow{p} X_1 \psi_1 \cdots X_m \psi_m$ is a stochastic production in P_s where $\psi_i = \lambda$ for $X_i \in V_T$. Then with probability

† See Fu and Huang [9].

p the string $\omega = \alpha A \theta \beta$, where α, $\beta \in (V_N F^* \cup V_T)^*$ and $\theta \in F^*$, directly generates the string $\eta = \alpha X_1 \varphi_1 \cdots X_m \varphi_m \beta$, where

$$\varphi_i = \begin{cases} \psi_i \theta & \text{if } X_i \in V_N \\ \lambda & \text{if } X_i \in V_T \end{cases}$$

for $i = 1, \ldots, k$.

(ii)   Suppose that $A \xrightarrow{p} X_1, \ldots, X_k$ is a stochastic indexed production in $f \in F$. Then, with probability p the string $\omega = \alpha A f \theta \beta$ directly generates the string $\eta = \alpha X_1 \varphi_1 \cdots X_m \varphi_m \beta$ in which

$$\varphi_i = \begin{cases} \theta & \text{if } X_i \in V_N \\ \lambda & \text{if } X_i \in V_T \end{cases}$$

for all $i = 1, \ldots, k$.

Definition 5.20   The stochastic indexed grammar operating under leftmost derivation is one for which the leftmost nonterminal in the intermediate string is always being expanded first in a derivation.

The stochastic language generated by a stochastic indexed grammar $G_{si} = (V_N, V_T, F, P_s, S)$ is

$$L(G_{si}) = \left\{ (x, p(x)) \mid x \in V_T^*, S \xrightarrow[*]{p_i} x, i = 1, \ldots, k, \text{ and } p(x) = \sum_{i=1}^{k} p_i \right\}$$

$$(5.21)$$

where k is the number of all distinctively different leftmost derivations of x from S. $L(G_{si})$ is called the stochastic indexed language. If

$$\sum_{x \in L(G_{si})} p(x) = 1$$

G_{si} is called a consistent stochastic indexed grammar. The consistency condition for a stochastic indexed grammar has yet to be found.

Example 5.10   Consider the stochastic indexed grammar $G_{si} = (V_N, V_T, F, P_s, S)$, where $V_N = \{S, T, A, B\}$, $V_T = \{a, b\}$, $F = \{f, g\}$, and P_s:

$$S \xrightarrow{1} Tf$$

$$T \xrightarrow{p} Tg \qquad (0 < p < 1)$$

$$T \xrightarrow{1-p} ABA$$

$$f: \quad A \xrightarrow{1} a \qquad g: \quad A \xrightarrow{1} aA$$

$$B \xrightarrow{1} b \qquad \qquad B \xrightarrow{1} bB$$

A typical derivation will be

$$S \Rightarrow Tf \Rightarrow Tgf \Rightarrow Agf\,Bgf\,Agf \Rightarrow aAf\,bBfaAf \Rightarrow aabbaa$$

and $p(aabbaa) = p(1 - p)$. In general,

$$L(G_{si}) = \{(a^n b^n a^n,\ p^{n-1}(1 - p)) | n = 1, 2, \ldots\}$$

which is a stochastic context-sensitive language.

5.5 Stochastic Syntax Analysis for Stochastic Context-Free Languages

The production probabilities in a stochastic grammar provide information about how often each production is used (when it is applicable) in the generation of the language. This information should be useful when we perform a parsing of a sentence generated by a stochastic grammar.† For example, suppose that, at a particular step of a top-down parsing process, there are two productions available for replacement. Intuitively speaking, we would apply the production with higher probability first if we want to reduce the possibility of committing a false trial. The use of the probability information in syntax analysis is called "stochastic syntax analysis" or "stochastic parsing." A block diagram for a system of stochastic syntax analysis is shown in Fig. 5.1.

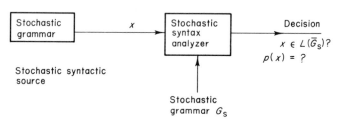

Fig. 5.1. Block diagram of a system for stochastic syntax analysis.

After a sentence x is accepted by the stochastic syntax analyzer, that is, $x \in L(\bar{G}_s)$, the probability of x being generated by G_s, $p(x)$, can be readily determined. If $x \notin L(\bar{G}_s)$, then $p(x) = 0$. In this section, algorithms for stochastic syntax analysis of context-free languages are presented. We have demonstrated that the use of stochastic syntax analysis algorithms would speed up the parsing in terms of the average number of analysis steps.

A bottom-up algorithm using precedence for the stochastic parsing of (stochastic) context-free languages has been proposed by Lee and Fu [24]. It

---

† If the grammar is a stochastic finite-state grammar, a stochastic finite automaton can be constructed to recognize the language generated by the grammar (Section 5.3.1).

has been shown by Fischer [25] and Learner and Lim [26] that every context-free grammar G_1 can be converted into a Wirth–Weber precedence grammar G_2 where $L(G_1) = L(G_2)$. This can also be shown to be true for a stochastic context-free language and its generating stochastic context-free grammar.

Lemma 5.1 Every stochastic context-free language L with consistent probability assignments has a Wirth–Weber stochastic precedence grammar [27] with consistent probability assignments.

The proof is similar to the one by Fischer [25] in which the grammar and language are nonstochastic. Let the grammar $G_s = (V_N, V_T, P_s, S)$ generate L, so $L(\bar{G}_s) = L$. Every production in G can be converted into Chomsky normal form. We get

$$(1) \quad A \xrightarrow{p} BC \qquad (2) \quad A \xrightarrow{p} a$$

where $A, B, C \in V_N; a \in V_T$. Then G is converted into G' by the introduction of a set of new nonterminals $\hat{V}_N$ to eliminate precedence conflicts.

$$G_s' = (V_N'; V_T, P_s', S), \qquad V_N' = V_N \cup \hat{V}_N, \qquad \hat{V}_N = \{\hat{A} \,|\, A \in V_N\}$$

The rules of P_s in grammar G_s are converted into rules in P_s' in grammar G_s' by the following:

(1)　For every rule $A \xrightarrow{p} BC$ in P_s with $A, B, C \in V_N$, P_s' has the rules $A \xrightarrow{p} \hat{B}C$ and $\hat{A} \xrightarrow{p} \hat{B}C$ for $\hat{A}, \hat{B} \in \hat{V}_N$ and $A, B, C \in V_N$.

(2)　For every rule of the form $A \xrightarrow{p} a$ in P_s, $a \in V_T$, P_s' has the rules $A \xrightarrow{p} a$ and $\hat{A} \xrightarrow{p} a$ for $a \in V_T$, $A \in V_N$, and $\hat{A} \in \hat{V}_N$. G_s' is a Wirth–Weber precedence grammar. If any precedence relation exists between symbols X and Y, it is given in the following, with $X, Y \in (V_N \cup \hat{V}_N \cup V_T)$.

	Y		
X	V_N	$\hat{V}_N$	V_T
V_N	$>$	$>$	$>$
$\hat{V}_N$	$\doteq$	$<$	$\lessdot$
V_T	$>$	$>$	$>$

Next we must prove that the consistency conditions are still valid in G_s'. Let $x \in L(G_s)$. Its derivation is

$$S \xRightarrow{r_1} \gamma_1 \xRightarrow{r_2} \gamma_2 \xRightarrow{r_3} \cdots \xRightarrow{r_n} \gamma_n = x$$

and

$$P(x) = \prod_{i=1}^{n} P(r_i)$$

If $A \xrightarrow{p(r_i)} BC$ was used in the original derivations at the ith step, then either $\hat{A} \xrightarrow{p(r_i)} \hat{B}C$ or $A \xrightarrow{p(r_i)} \hat{B}C$ is used in grammar G_s' derivation. Also, if $A \xrightarrow{p(r_j)} a$ was used originally in the jth step, then either $\hat{A} \xrightarrow{p(r_j)} a$ or $A \xrightarrow{p(r_j)} a$ is used in grammar G_s'. Therefore,

$$p'(x) = \prod_{i=1}^{n} p(r_i) \quad \text{and} \quad p'(x) = p(x)$$

A stochastic context-free grammar may contain many productions whose right parts are identical. When parsing using this grammar, the parsing algorithm becomes nondeterministic; that is, there may be many alternatives for the reduction of a handle. All possibilities must be tried, until either the string is accepted, or no alternatives are left and the string is rejected. However, the probability information of the language can be used to reduce the parsing time [24]. A syntax-controlling probability (scp) is calculated for each production. Associated with each production $A_j \xrightarrow{p_{ji}} \eta_i$ (p_{ji} is the normal production probability), the scp $\tilde{p}_{ji}$ is defined as probability that A_i is the correct nonterminal to be used in a replacement of η_i. This probability is calculated from the expected number of times that $A_j \to \eta_i$ is used in the overall generation of all strings of $L(G_s)$, with respect to all the productions with the same right part. Therefore, it can be shown that the scp indicates the most probable production to apply.

All the productions are divided into equivalent classes, $\Gamma_{\eta_1} \cdots \Gamma_{\eta_n}$, where all members of a class Γ_{η_1} have identical right parts.

$$A_1 \xrightarrow[\tilde{p}_{11}]{p_{11}} \eta_1 \cdots A_q \xrightarrow[\tilde{p}_{qn}]{p_{qn}} \eta_n$$

$$\vdots \qquad\qquad \vdots$$

$$A_r \xrightarrow[\tilde{p}_{r1}]{p_{r1}} \eta_1 \cdots A_m \xrightarrow[\tilde{p}_{mn}]{p_{mn}} \eta_n$$

where all $A_j \in V_N$ and all $\eta_i \in V^+$. The scp's are normalized with respect to the right-part equivalent class.

$$\sum_{A_j \to \eta_i \in \Gamma_{\eta_i}} \tilde{p}_{ji} = 1 \tag{5.22}$$

where the summation is over all productions in Γ_{η_i}.

Let $N_{ji}(x)$ be the number of times that the production $A_j \to \eta_i$ (member of equivalent class Γ_{η_i}) is used in the derivation of the string x. The probability $\tilde{p}_{ji}$ can be computed from the chance of occurrence of production $A_j \to \eta_i$ divided by the total occurrence of all Γ_{η_i} productions, so

$$\tilde{p}_{ji} = \frac{\sum_{x \in L} p(x) N_{ji}(x)}{\sum_{A_j \to \eta_i \in \Gamma_{\eta_i}} \sum_{x \in L} p(x) N_{ji}(x)} \tag{5.23}$$

The double summation in the denominator is over all the strings in $L(\bar{G}_s)$ and all productions in Γ_{η_i}. Equation (5.23) represents the exact value of $\tilde{p}_{ji}$ where $p(x)$ is the probability of generating string x and $0 < \tilde{p}_{ji} \le 1$. Using Eq. (5.23), the scp can be found. However, all the probabilities of the strings $x \in L(\bar{G}_s)$ must be known.

Since a context-free language can be modeled by a multiple-type Galton–Watson branching process, we can also use this process to describe the $\tilde{p}_{ji}$ probabilities. The nonterminals are the states of the process. Let N_{ji} be the expected number of times production $A_j \to \eta_i$ is used in the average or expected parse. N_{ji} is expressed as the product of a conditional expectation and the production probability of $A_j \to \eta_i$, which is p_{ji}. The conditional expectation is the expected number of times that the process is in state A_j at any branch level, given that it started in state A_0, where A_0 is the start symbol. The conditional expectations at the nth branch level expressed in matrix form are

$$E(\mathbf{A}_n | \mathbf{A}_0) = \mathbf{A}_0 E^n \tag{5.24}$$

where E is the matrix of the first moments and $\mathbf{A}_n$ and $\mathbf{A}_0$ are the state vectors of the branch levels 0 and n, respectively. E is obtained from the generating function for the multitype Galton–Watson process (Section 5.2.2). Let $f_i(s_1, s_2, \ldots, s_n)$ be the generating function for the ith nonterminal, where V_N consists of n nonterminals. E is defined by (5.10) and (5.9). Then

$$N_{ji} = \sum_{n=0}^{\infty} [E(\mathbf{A}_n | \mathbf{A}_0)]\xi_j \, p_{ji} \tag{5.25}$$

$\mathbf{A}_0$ is a row vector with its first element equal to 1 and ξ_j is a column vector with the jth element equal to 1. So N_{ji} equals the sum of all branch levels from 0 to ∞ that start in state j and go into state i. Substituting Eq. (5.24) into Eq. (5.25) gives

$$N_{ji} = \left[\sum_{n=0}^{\infty} \mathbf{A}_0 E^n \right] \xi_j \, p_{ji}$$

If the grammar is consistent, then the largest absolute value of the eigenvalue is less than 1, and $\lim_{n \to \infty} E^n = 0$. Let

$$\tilde{B}_{0j} = \mathbf{A}_0 \left[\sum_{n=0}^{\infty} E^n \right] \xi_j = A_0 (I - E)^{-1} \xi_j \tag{5.26}$$

so $N_{ji} = \tilde{B}_{0j} \, p_{ji}$. To normalize N_{ji} we divide it by the sum of all the N_{ji}'s of the equivalent class, Γ_{η_i}; that is,

$$\tilde{p}_{ji} = \frac{\tilde{B}_{0j} \, p_{ji}}{\sum_{A_k \to \eta_i \in \Gamma_{\eta_i}} \tilde{B}_{0k} \, p_{ki}} \tag{5.27}$$

Again the summation in the denominator is over all productions in Γ_{η_i}.

Parsing time using the scp in a stochastic parser can be shown to be equal to or faster than that when using a corresponding nondeterministic parser (without using the probability information). Speed or time savings will be measured in terms of the expected number of steps. In order to make a quantitative measure of the expected number of steps of a nondeterministic parser, the following measure will be used. All possible right-part ordering configurations are considered equally likely. Therefore, the average of all of these configurations will be used to calculate the expected number of steps of a nondeterministic parser. The number of error steps and backtracking steps are used to make a meaningful speed comparison; the nondeterministic parser and the stochastic parser use the same amount of forward steps for any parse.

Grammar $G_s = (V_N, V_T, P_s, A_1)$ must obey the following conditions:

(1)   The grammar G_s is consistent.
(2)   The grammar G_s is in Wirth–Weber precedence form.
(3)   The grammar G_s is unambiguous.
(4) λ_i is the expected number of steps that take place before an error is recognized and the expected number of steps that are backtracked to the position where the next right part can be applied. λ_i is the same for any one of the identical right parts in the equivalent class Γ_{η_i}.

Condition (4) is needed in order to obtain a closed form measure of the wasted or extra steps. Condition (3) must hold for the following reason. If the grammar is ambiguous, all possible replacements must be tried to be sure no other parse exists. If all possible replacements are tried, then no savings in steps can be made.

As a consequence of these four conditions we have a precedence grammar and a λ_i associated with each right-part equivalent class Γ_{η_i}.

Production Number	Productions
1	$A_{j_1} \xrightarrow{p_{j_1 i}} \eta_i$
2	$A_{j_2} \xrightarrow{p_{j_2 i}} \eta_i$
$\vdots$	
m	$A_{j_m} \xrightarrow{p_{j_m i}} \eta_i$

If there are m productions in a right-part equivalent class Γ_{η_i}, then there are $m!$ possible orderings. Let a subscript j_k represent the kth term in the jth ordering. Also, let $j = l$ be the ordering induced by the scp.

$$N_{l_1 i} \geq N_{l_2 i} \geq \cdots \geq N_{l_m i} \tag{5.28}$$

Theorem 5.11   Given conditions (1)–(4), then

$$E\{n_{\text{sp}}\} \leq E\{n_{\text{np}}\} \tag{5.29}$$

where $E\{n_{\text{sp}}\}$ is the expected number of steps of the stochastic parser and $E\{n_{\text{np}}\}$ is the expected number of steps of the nondeterministic parser.

Proof.   Let $E\{n_{\text{f}}\}$ be the expected number of forward steps, which is the same for both parsers. The expected number of error steps incurred by the hth ordering of the right-part equivalent class Γ_{n_i} is calculated by

$$E\{n_h{}^i\} = \lambda_i \sum_{k=2}^{m} (k-1)N_{h_k i} \tag{5.30}$$

where λ_i is the number of steps incurred before selecting a new identical right part. If the first right part selected was correct, no errors are made; therefore, $N_{h_1 i}$ does not appear in $E\{n_h{}^i\}$. If the second production was correct, then λ_i errors occurred for each production so used. If the kth production was the correct production, then $(k-1)\lambda_i$ errors occurred for each production used in that manner.

$E\{n_h{}^i\}$ is a minimum for the case when $N_{h_1 i} \geq N_{h_2 i} \geq \cdots \geq N_{h_m i}$. This is precisely the case for the scp ordering.

$$E\{n_h{}^i\}_{\min} = \lambda_i \sum_{k=2}^{m} (k-1)N_{l_k i} = E\{n_{\text{sp}}^i\} \tag{5.31}$$

where the minimum is over all the $m!$ orderings. For any other ordering,

$$E\{n_{\text{sp}}^i\} \leq E\{n_h{}^i\}, \qquad 1 < h \leq m! \tag{5.32}$$

Summing over all possible orderings, we have

$$\sum_{l=1}^{m!} E\{n_{\text{sp}}^i\} \leq \lambda_i \sum_{l=1}^{m!} \sum_{k=2}^{m} (k-1)N_{l_k i}$$

$$E\{n_{\text{sp}}^i\} \leq \frac{\lambda_i}{m!} \sum_{l=1}^{m!} \sum_{k=2}^{m} (k-1)N_{l_k i} = E\{n_{\text{np}}^i\} \tag{5.33}$$

$$E\{n_{\text{sp}}^i\} \leq E\{n_{\text{np}}^i\}$$

The right-hand side of inequality (5.33) is the definition of the expected number of error steps of the nondeterministic parser.

Summing $E\{n_{\text{sp}}^i\}$ and $E\{n_{\text{np}}^i\}$ over all equivalent classes and adding $E\{n_{\text{f}}\}$ yields the conclusion desired.

$$E\{n_{\text{f}}\} + \sum_{i=1}^{n} E\{n_{\text{sp}}^i\} \leq \sum_{i=1}^{n} E\{n_{\text{np}}^i\} + E\{n_{\text{f}}\}, \qquad E\{n_{\text{sp}}\} \leq E\{n_{\text{np}}\} \tag{5.34}$$

Condition (4) was made in order to make it possible to obtain a quantitative comparison of the number of error steps for a right-part equivalence class.

This condition is always true for the following situation. Let $A_{j_1}, A_{j_2}, \ldots, A_{j_m}$ be the m left-hand nonterminals of the right-part equivalence class Γ_{n_i}. If A_{j_k} $(1 \le k \le m)$ appears in sentential form $\alpha X_l A_{j_k} X_{l+1} \gamma$, where α, $\gamma \in (V_T \cup V_N)^*$ and $X_l, X_{l+1} \in (V_N \cup V_T \cup \lambda)$, then A_{j_q} $(1 \le q \le m$ and $q \ne k)$ does not appear in the sentential form $\delta X_l A_{j_q} X_{l+1} \beta$ for any $\delta, \beta \in (V_T \cup V_N)^*$. In this case the errors are discovered in the precedence matrix. Here $\lambda_i = 2$ (one error step and one backtrack step).

Example 5.11 An example of finding the scp for a linear grammar with multiple identical right parts is as follows.

$$G_s = (V_N; V_T, P_s, S), \qquad V_N = \{S, A, B\}, \qquad V_T = \{a, b, c, d\}$$

$$P_s: \quad S \xrightarrow{p} aSb \qquad A \xrightarrow{1-q} Bd$$

$$S \xrightarrow{1-p} Ac \qquad B \xrightarrow{r} aSb$$

$$A \xrightarrow{q} aSb \qquad B \xrightarrow{1-r} d$$

	S	A	B	a	b	c	d
S					$\doteq$		
A						$\doteq$	
B							$\doteq$
a	$\doteq$	$\lessdot$	$\lessdot$	$\lessdot$			
b					$\gtrdot$	$\gtrdot$	$\gtrdot$
c					$\gtrdot$		
d					$\gtrdot$	$\gtrdot$	

We want to find all the $\bar{p}$'s for the grammar G_s.

Production numbers are ordered according to their scp for the minimum error.

(1) $S \xrightarrow[\bar{p}_{SS}=0.704]{p=0.7} aSb \qquad \bar{p}_{SS} = \dfrac{p}{p + q(1-p) + r(1-p)(1-q)}$

(2) $A \xrightarrow[\bar{p}_{AS}=0.242]{q=0.8} aSb \qquad \bar{p}_{AS} = \dfrac{q(1-p)}{p + q(1-p) + r(1-p)(1-q)}$

(3) $B \xrightarrow[\bar{p}_{BS}=0.054]{r=0.9} aSb \qquad \bar{p}_{BS} = \dfrac{r(1-p)(1-q)}{p + q(1-p) + r(1-p)(1-q)}$

The scp's for the rest of the productions are equal to 1. For this right-part equivalence class all errors are detected in the precedence matrix, so $\lambda = 2$

(1 step for the error and 1 step for the backtrack). There are six possible orderings. The average number of error steps for the stochastic parser is 116. The other five orderings used 178, 270, 394, 486, and 548 expected number of error steps, respectively. So the average number of error steps for the nondeterministic parser is 332. The average number of forward steps is 227. The average string length is 393. Therefore, the average number of analysis steps for the nondeterministic parser is 559, whereas for the stochastic parser it is 343. The stochastic parser has been applied to the problem of classifying ambiguous chromosome patterns [24, 28].

5.6   Stochastic Syntax Analysis for Context-Free Programmed Languages

In this section we will present two basic approaches to the construction of a stochastic syntax analyzer which is used to determine whether a string x can be generated by a given scfpg G_{sp} and to compute the probability of the string x being generated by G_{sp} [29]. The stochastic syntax analyzer is basically a top-down parsing algorithm consisting of two basic parts: (1) a stochastic generating algorithm, and (2) a deterministic backtracking algorithm. Depending on the type of stochastic selection algorithm used, we have two distinct types of stochastic syntax analyzers. The first type is one with a selection algorithm which searches through the most likely production first, while the second type is one with a selection algorithm which randomly selects a production in a go-to field according to the distribution over all possible alternatives defined in the go-to field.

We shall illustrate by examples that the stochastic syntax analyzer for a scfpg G_{sp} that searches through the most likely alternative production first would speed up the analysis in the sense of reducing the average number of steps required to recognize the language $L(\bar{G}_{sp})$. We shall also show that the stochastic syntax analyzer which randomly selects an alternative production according to the probability distribution given in the grammar may also improve the performance of the syntax analyzer in the sense of reducing the average number of steps required to recognize the language $L(\bar{G}_{sp})$.

5.6.1   Stochastic Syntax Analyzer with a Fixed Strategy

The stochastic syntax analyzer that searches the most likely production first is a nondeterministic syntax analyzer [8] in which the elements contained in each go-to field are arranged in descending order of magnitude of their associated probabilities. The leftmost (or the first) production label in each go-to field is the one corresponding to the highest probability, while the rightmost (or the last) production label in each go-to field is the one corresponding to the lowest probability. Clearly, this is a stochastic syntax analyzer

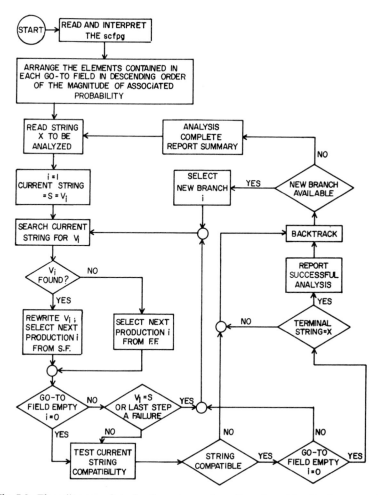

Fig. 5.2. Flow diagram of stochastic syntax analyzer for scfpg with a fixed strategy (S.F. means success field, F.F., failure field).

with a fixed strategy. A flow diagram for this stochastic syntax analyzer with a fixed strategy is shown in Fig. 5.2.

We will show by the following example that the stochastic syntax analyzer which searches through the most likely production first would require the least average number of steps to recognize the language generated by a stochastic cfpg.

Example 5.12   Consider a stochastic cfpg $G_1 = (\{S, A, B, C\}, \{a, b, c\}, \{1, \ldots, 7\}, P_s, S)$ where P_s is defined as follows:

Label	Core	U	$P(U)$
1	$S \rightarrow ABC$	2, 5	$p, 1 - p$
2	$A \rightarrow aA$	3	1
3	$B \rightarrow bB$	4	1
4	$C \rightarrow cC$	2, 5	$q, 1 - q$
5	$A \rightarrow a$	6	1
6	$B \rightarrow b$	7	1
7	$C \rightarrow c$	$\varnothing$	1

in which $0 < p, q < 1$. It is a simple matter to show that the string abc can be derived from S according to a derivation $r_1 r_5 r_6 r_7$ with probability

$$p(abc) = p(r_1)p(r_5|r_1)p(r_6|r_5)p(r_7|r_6) = (1)(1 - p)(1)(1) = 1 - p$$

In general, the string $a^n b^n c^n$, for $n > 1$, can be derived from S according to a derivation $r_1(r_2 r_3 r_4)^{n-1} r_5 r_6 r_7$ with probability

$$p(a^n b^n c^n) = p(r_1)p(r_2|r_1)p(r_3|r_2)p(r_4|r_3)[p(r_2|r_4)$$
$$p(r_3|r_2)p(r_4|r_3)]^{n-2}p(r_5|r_4)p(r_6|r_5)p(r_7|r_6)$$
$$= (1)(p)(1)(1)[(q)(1)(1)]^{n-2}(1 - q)(1)(1) = p(1 - q)q^{n-2}$$

In summary, the stochastic language generated by G_1 is

$$L(G_1) = \{(a^n b^n c^n, p_n)|n > 0\}$$

where

$$p_1 = 1 - p \quad \text{and} \quad p_n = p(1 - q)q^{n-2}, \quad n = 2, 3, \ldots$$

G_1 can be considered as a stochastic source and the string $a^n b^n c^n$ is generated with probability p_n.

We shall now calculate the average number of steps required by the syntax analyzer, which are obtained from G_1 by reordering the elements in the go-to fields to recognize the language $L(\bar{G}_1)$. For the scfpg G_1, there are only four possible different ways to order the production labels in the go-to fields associated with the first and the fourth productions. Based on the order of all alternatives in a go-to field, G_2, G_3, and G_4 are formed from G_1 by rearranging the order of all production labels in the go-to fields in G_1. We also assume that the order of all alternatives in the go-to fields in a scfpg G_{sp} is the order that a syntax analyzer constructed according to G_{sp} will search through during the syntax analysis. We shall now construct G_2, G_3, and G_4 from G_1 and calculate the average number of steps required by each of four syntax analyzers constructed according to G_1, G_2, G_3, and G_4 to recognize the language $L(\bar{G}_1)$.

(1)   If the syntax analyzer is constructed according to G_1, the number of steps required by the analyzer to recognize $a^n b^n c^n$ is $3n + 2$. Hence, the average number of steps required by the syntax analyzer to recognize $L(\bar{G}_1)$, denoted by D_1, is

$$D_1 = \sum_{n=1}^{\infty} (3n + 2)p_n = 5(1 - p) + \sum_{n=2}^{\infty} (3n + 2)P(1 - q)q^{n-2}$$

$$= 5(1 - p) + p(1 - q)\sum_{m=0}^{\infty} (3m + 8)q^m$$

$$= 5(1 - p) + p(1 - q)\left[3\sum_{m=0}^{\infty} mq^m + 8\sum_{m=0}^{\infty} q^m\right]$$

$$= 5(1 - p) + p(1 - q)[3q/(1 - q)^2 + 8/(1 - q)]$$

$$= 5 + 3p/(1 - q) \tag{5.35}$$

(2)   The syntax analyzer is constructed according to the grammar G_2 whose production set is defined as follows:

Label	Core	U	P(U)
1	$S \rightarrow ABC$	2, 5	$p, (1 - p)$
2	$A \rightarrow aA$	3	1
3	$B \rightarrow bB$	4	1
4	$C \rightarrow cC$	5, 2	$(1 - q), q$
5	$A \rightarrow a$	6	1
6	$B \rightarrow b$	7	1
7	$C \rightarrow c$	$\varnothing$	1

It is noted that G_2 is formed from G_1 by reordering the elements contained in the success field in the fourth production. Clearly, $L(G_1) = L(G_2)$.

It can be shown that the number of steps required by this syntax analyzer to recognize the string abc is 5. It takes $3n + 2(n - 2)$ steps to recognize $a^n b^n c^n$, for $n > 1$. Hence, the average number of steps required by the analyzer to recognize $L(\bar{G}_1)$, denoted by D_2, is

$$D_2 = 5(1 - p) + \sum_{n=2}^{\infty} [3n + 2(n - 2)]p(1 - q)q^{n-2}$$

$$= 5 + 3p/(1 - q) + 2p(2q - 1)/(1 - q) \tag{5.36}$$

(3)   The syntax analyzer is constructed according to the grammar G_3 whose production set is defined as follows:

Label	Core	U	$P(U)$
1	$S \to ABC$	5, 2	$(1-p), p$
2	$A \to aA$	3	1
3	$B \to bB$	4	1
4	$C \to cC$	2, 5	$q, (1-q)$
5	$A \to a$	6	1
6	$B \to b$	7	1
7	$C \to c$	$\varnothing$	1

It is noted that G_3 is formed from G_1 by reordering the elements contained in the success field of the first production. Clearly, $L(G_3) = L(G_1)$.

It can be shown that it takes 3 steps to recognize the string abc and $3n + 4$ to recognize the string $a^n b^n c^n$, for $n > 1$. Hence, the average number of steps required by this analyzer to recognize $L(\bar{G}_1)$, denoted by D_3, is

$$D_3 = 3(1 - p) + \sum_{n=2}^{\infty} (3n + 4)p(1 - q)q^{n-2} = 5 + 3p/(1 - q) + (4p - 2)$$

$$(5.37)$$

(4)   Finally, the stochastic syntax analyzer is constructed according to the grammar G_4, whose production set is defined as follows:

Label	Core	U	$P(U)$
1	$S \to ABC$	5, 2	$(1-p), p$
2	$A \to aA$	3	1
3	$B \to bB$	4	1
4	$C \to cC$	5, 2	$(1-q), q$
5	$A \to a$	6	1
6	$B \to b$	7	1
7	$C \to c$	$\varnothing$	1

It is noted that G_4 is formed from G_1 by reordering the elements contained in the success fields in the first and fourth productions. Clearly, $L(G_4) = L(G_1)$.

It can be shown that it takes 3 steps to recognize the string abc and $3n + 2(n - 1)$ steps to recognize $a^n b^n c^n$, $n > 1$. Hence, the average number of steps required by the analyzer to recognize $L(\bar{G}_1)$, denoted by D_4, is

$$D_4 = 3(1 - p) + \sum_{n=2}^{\infty} [3n + 2(n - 1)]p(1 - q)q^{n-2} = 3 + 5p/(1 - q) \quad (5.38)$$

In summary, for $0 < p, q < 1$,

$$D_1 = 5 + 3p/(1 - q), \qquad\qquad D_3 = 3 + 3p/(1 - q) + 4p$$
$$D_2 = 5 + 3p/(1 - q) + 2p(2q - 1)/(1 - q), \qquad D_4 = 3 + 5p/(1 - q)$$

Let

$$D = \min (D_1, D_2, D_3, D_4)$$

It can be shown easily that [29]:

$$D = \begin{cases} D_1 \text{ or } D_2 \text{ or } D_3 \text{ or } D_4 & \text{if } p = \tfrac{1}{2}, \ q = \tfrac{1}{2} \\ D_2 \text{ or } D_4 & \text{if } p = \tfrac{1}{2}, \ q < \tfrac{1}{2} \\ D_1 \text{ or } D_3 & \text{if } p = \tfrac{1}{2}, \ q > \tfrac{1}{2} \\ D_3 \text{ or } D_4 & \text{if } p < \tfrac{1}{2}, \ q = \tfrac{1}{2} \\ D_1 \text{ or } D_2 & \text{if } p > \tfrac{1}{2}, \ q = \tfrac{1}{2} \\ D_1 & \text{if } p > \tfrac{1}{2}, \ q > \tfrac{1}{2} \\ D_2 & \text{if } p > \tfrac{1}{2}, \ q < \tfrac{1}{2} \\ D_3 & \text{if } p < \tfrac{1}{2}, \ q > \tfrac{1}{2} \\ D_4 & \text{if } p < \tfrac{1}{2}, \ q < \tfrac{1}{2} \end{cases} \qquad (5.39)$$

The results of Example 5.12 clearly illustrate that the stochastic syntax analyzer which searches through the most likely production first would require the least average number of steps to recognize the language $L(\bar{G}_1)$.

5.6.2 *Stochastic Syntax Analyzer with a Random Strategy*

The stochastic syntax analyzer with a random strategy is formed from the nondeterministic syntax analyzer by incorporating a stochastic selection algorithm in selecting the next production in a given go-to field when alternatives are available. At each step the conditional probability distribution is used in selecting the next production among all available alternatives. This notion can be clearly illustrated by the following simple example. Suppose that

Label	Core	U	$P(U)$
r	$A \to aB$	r_i, r_j, r_k	p_i, p_j, p_k

where $p_i + p_j + p_k = 1$ is a production successfully applied to the current intermediate string. The next production is statistically selected among r_i, r_j, and r_k with associated probabilities p_i, p_j, and p_k, respectively. Suppose that r_j is selected as the next production and leads to no successful analysis of the given string. The backtracking algorithm will backtrack to

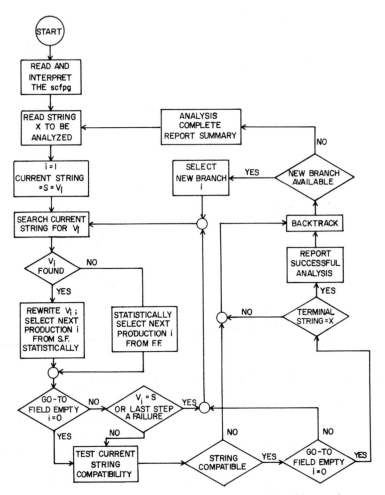

Fig. 5.3. Flow diagram of stochastic syntax analyzer for scfpg with a random strategy.

production r. Since r_j has been previously applied, r_i and r_k are the only available productions in U. Hence, either r_i or r_j can be selected, according to their conditional probability distribution $P(r_i, r_k | r_j) = (p_i', p_k')$ where $p_i' = p_i/(p_i + p_k)$ and $p_k' = p_k/(p_i + p_k)$. A flow diagram for the stochastic syntactic analyzer of the second type is shown in Fig. 5.3.

It is easy to see that a stochastic syntax analyzer with a random strategy is equivalent to a nondeterministic syntax analyzer with mixed strategies. That is, the stochastic syntax analyzer with a random strategy can be considered as a nondeterministic syntax analyzer which employs a finite set of fixed strategies, each of which is selected by the nondeterministic analyzer to be the strategy with some probability. It is also evident that the probability of selecting some strategy is equal to 1. We shall show that the probability of a particular fixed strategy's being selected by the nondeterministic syntax analyzer can be calculated from the given scfpg under consideration.

Suppose that for a given scfpg G_{sp} there are N fixed strategies (or distinctly different orderings for elements contained in the go-to fields in G_{sp}) and π_i is the probability that the ith strategy is used by the nondeterministic analyzer with mixed strategies. Then the average number of steps required by the non-deterministic analyzer with mixed strategies to recognize $L(\bar{G}_{sp})$, denoted by D', is

$$D' = \sum_{i=1}^{N} \pi_i D_i \tag{5.40}$$

where D_i is the average number of steps required by the nondeterministic syntax analyzer with the ith strategy to recognize the language $L(\bar{G}_{sp})$, and

$$\sum_{i=1}^{N} \pi_i = 1$$

It is now easy to see that

$$\max_i D_i \geq D' \geq \min_i D_i \tag{5.41}$$

In other words, the stochastic syntax analyzer with a random strategy may be more efficient than the nondeterministic syntax analyzer with some fixed strategy but is less efficient than the stochastic syntax analyzer with a fixed strategy.

For a given scfpg G_{sp}, the order of elements contained in each go-to field in G_{sp} defines a fixed strategy for the nondeterministic syntax analyzer with mixed strategies. The probability π_i of using the ith strategy can be calculated from G_i, which corresponds to the ith strategy. Suppose that

$$r \quad A \to \omega \quad U = \{r_1, \dots, r_k\} \quad P(U) = \{p_1, \dots, p_k\}$$

where

$$\sum_{i=1}^{k} p_i = 1$$

is a production with a go-to field that contains k elements. The probability associated with the ordering of the k elements contained in the go-to field of the rth production, denoted by $P(r)$, is

$$P(r) = \prod_{j=1}^{k} p_j \Big/ \left(\sum_{i=j}^{k} P_i \right) \tag{5.42}$$

in which $p_j/(\sum_{i=j}^{k} p_i)$ is the conditional probability that r_j is the jth element in the go-to field of the rth production given that $r_1, \dots, r_{j-1}$ are the first $(j-1)$

elements in the go-to field. The probability associated with the ith strategy of G_i is

$$\pi_i = \prod_{i \in J} P(r)$$

where J is the set of production labels defined in G_i. With some manipulation, it can be shown that

$$\sum_{i=1}^{N} \pi_i = 1$$

where N is the number of all possible different strategies defined by a scfpg G_{sp} by properly ordering the elements contained in each go-to field in G_{sp}. The behavior of a stochastic syntax analyzer with a random strategy can be illustrated by the following example.

Example 5.13   For the stochastic cfpg considered in Example 5.12, it can be shown that the stochastic syntax analyzer with a random strategy will take $(3 + 2p)$, $(6 + 2q + 2(1 - p))$, and $(3n + 2 + 2(n - 3)(1 - q) + 2(1 - p))$ steps to recognize abc, $a^2b^2c^2$, and $a^nb^nc^n$, $n \geq 3$, respectively. The average number of steps required by this stochastic analyzer with a random strategy to recognize $L(\bar{G}_1)$, denoted by D^*, is

$$D^* = (3 + 2p)(1 - p) + (6 + 2q + 2(1 - p))p(1 - q)$$

$$+ \sum_{n=3}^{\infty} [(3n + 2) + 2(n - 3)(1 - q) + 2(1 - p)]p(1 - q)q^{n-2}$$

$$= 3 + 3p/(1 - q) + 4p(1 - p) + 4pq \tag{5.43}$$

For this example, the stochastic syntax analyzer with a random strategy can be also considered as a nondeterministic syntax analyzer with mixed strategies. It has been shown that there are four strategies to be used by the nondeterministic syntax analyzer: G_1, G_2, G_3, and G_4, defined in Example 5.12. The probability π_i of choosing the ith strategy can be calculated from the scfpg G_i. It can be shown that π_1, π_2, π_3, and π_4 are equal to pq, $p(1 - q)$, $(1 - p)q$, and $(1 - p)(1 - q)$, respectively. Then the average number of steps required by the stochastic syntax analyzer with a random strategy to recognize the language $L(\bar{G}_1)$ can be simply calculated as follows.

$$D' = \sum_{i=1}^{4} \pi_i D_i$$

$$= pq(5 + 3p/(1 - q)) + p(1 - q)(5 + 3p/(1 - q) + 2q - 1)/(1 - q))$$

$$+ (1 - p)q(3 + 3p/(1 - q) + 4p) + (1 = p)(1 - q)(3 + 5p/(1 - q))$$

$$= 3 + 3p/(1 - q) + 4p(1 - p) + 4pq \tag{5.44}$$

which is the same as D^*.

It can be shown that $D = D^*$ for the case in which $p = q = \frac{1}{2}$, and that $D < D^*$ for all other cases [29].

In order to show that the stochastic syntax analyzer would speed up the parsing in the sense of reducing the average number of steps required to recognize the language $L(\bar{G}_1)$, we shall now show that $D^* \leq D_n$, in which D_n is defined as the mean of the average number of steps required by syntax analyzers constructed according to G_1, G_2, G_3, and G_4. By definition,

$$D_n = (D_1 + D_2 + D_3 + D_4)/4 = 4 + 4p/(1 - p) \qquad (5.45)$$

Then

$$D_n - D^* = (1 - 2p)^2 + (1 - 2q)^2/(1 - q) \geq 0$$

It is noted that $D_n = D^*$ if and only if $p = q = \frac{1}{2}$. Hence, we establish that

$$D \leq D^* \leq D_n \qquad (5.46)$$

in which $D = D^* = D_n$ if and only if $p = q = \frac{1}{2}$.

It is now clear that if the probability information associated with a scfpg G_{sp} is completely known, the stochastic syntax analyzer with a fixed strategy should be used for analysis. On the other hand, if the probability information associated with a scfpg is not completely known, then it might be better to use the stochastic syntax analyzer with a random strategy, since it is no worse than the nondeterministic syntax analyzer with the worst strategy. This situation may arise when the scfpg is inferred or heuristically constructed from a finite set of strings generated by an unknown scfpg.

References

1. R. B. Banerji, Phrase structure languages, finite machines, and channel capacity. *Information and Control* 6, 153–162 (1963).
2. U. Grenander, *Syntax-Controlled Probabilities.* Tech. Rep. Div. of Appl. Math., Brown Univ., Providence, Rhode Island, 1967.
3. K. S. Fu and P. H. Swain, On syntactic pattern recognition. *Int. Symp. Comput. and Inform. Sci., Bal Harbour, Florida, December 1969;* published in *Software Engineering* (J. T. Tou, ed.), Vol. II. Academic Press, New York, 1971.
4. K. S. Fu, Stochastic automata, stochastic languages and pattern recognition. *IEEE Symp. Decision and Contr., Austin, Texas, December 1970;* published in *J. Cybernet.* **1**, 31–49 (1971).
5. K. S. Fu, On syntactic pattern recognition and stochastic languages. In *Frontiers of Pattern Recognition* (S. Watanabe, ed.). Academic Press, New York, 1972.
6. S. E. Hutchins, *Stochastic Sources for Context-Free Languages.* Ph.D. Thesis, Dept. of Appl. Phys. and Inform. Sci., Univ. of California, San Diego, 1970.
7. C. R. Souza, *Probabilities in Context-Free Programmed Grammars.* TR-A70-2. Aloha Syst. Univ. of Hawaii, Honolulu, Hawaii, 1970.
8. P. H. Swain and K. S. Fu, Stochastic programmed grammars for syntactic pattern recognition. *Pattern Recognition* **4** (1972). Special issue on syntactic pattern recognition.

9. K. S. Fu and T. Huang, Stochastic grammars and languages. *Int. J. Comput. and Inform. Sci.* **1** (1972).
10. T. L. Booth, Probability Representation of Formal Languages. *IEEE Annu. Symp. Switching and Automata Theory, 10th, October 1969.*
11. T. E. Harris, *The Theory of Branching Processes.* Springer-Verlag, Berlin and New York, 1963.
12. B. A. Sevast'yanov, Theory of branching processes. *Progr. Math.* **7** (1970).
13. M. O. Rabin, Probabilistic automata. *Information and Control* **6**, 230–245 (1963).
14. A. Paz, Some aspects of probabilistic automata. *Information and Control* **9**, 26–60 (1966).
15. P. Turakainen, On stochastic languages. *Information and Control* **12**, 304–313 (1968).
16. A. Salomaa, On languages accepted by probabilistic and time-variant automata. *Proc. Annu. Conf. Inform. Sci. and Syst. 2nd, Princeton, 1968*, pp. 184–188.
17. K. S. Fu and T. Li, On stochastic automata and languages, *Information Sci.* **1**, 403–419 (1969).
18. C. A. Ellis, *Probabilistic Languages and Automata.* Rep. No. 355. Dept. of Comput. Sci., Univ. of Illinois, Urbana, 1969.
19. T. Huang and K. S. Fu, On stochastic context-free languages. *Information Sci.* **3**, 201–224 (1971).
20. M. P. Schutzenberger, On context-free languages and pushdown automata. *Information and Control* **3**, 246–264 (1963).
21. S. Ginsburg and S. A. Greibach, Deterministic context-free languages. *Information and Control* **9**, 620–648 (1966).
22. D. J. Rosenkrantz, Programmed grammars—A new device for generating formal languages. *J. Assoc. Comput. Mach.* **10**, 107–131 (1969).
23. A. V. Aho, Indexed grammars—An extension of context-free grammars. *J. Assoc. Comput. Mach.* **15**, 647–671 (1968).
24. H. C. Lee and K. S. Fu, A stochastic syntax analysis procedure and its application to pattern classification. *IEEE Trans. Computers* **C-21**, 660–666 (1972).
25. M. Fischer, Some properties of precedence languages. *Proc. ACM Symp. Theory of Comput. 1st, 1969*, pp. 181–190.
26. A. Learner and A. L. Lim, Note on transformating context-free grammars to Wirth-Weber precedence form. *Comput. J.* **13**, 142–144 (1970).
27. N. Wirth and H. Weber, EULER—A generalization of ALGOL and its formal definition, Pt. 1. *Comm. ACM* **9**, 13–25 (1966).
28. H. C. Lee and K. S. Fu, A syntactic pattern recognition with learning capability. *Int. Symp. Comput. Inform. Sci. 4th, Miami Beach, Florida, December 14–16, 1972* (J. T. Tou, ed.). Academic Press, New York, 1972.
29. T. Huang and K. S. Fu, Stochastic syntactic analysis for programmed grammars and syntactic pattern recognition. *Comput. Graphics and Image Process.* **1**, 257–283 (1972).
30. G. D. Bruce and K. S. Fu, A model for finite-state probabilistic systems. *Proc. Conf. Circuit and Syst. Theory, 1st, Allerton, Univ. of Illinois, 1962.*
31. M. M. Kherts, Entropy of languages generated by automata or context-free grammars with a single-valued deduction. *Naucho-Tekhnicheskaia Informatsia, Ser. 2, No. 1* (1968).
32. A. Salomaa, On m-adic probabilistic automata. *Information and Control* **10**, 215–219 (1967).
33. A. Salomaa, Probabilistic and weighted grammars. *Information and Control* **15**, 529–544 (1969).
34. E. S. Santos, Maximin automata. *Information and Control* **13**, 363–377 (1968).
35. P. Turakainen, Some closure properties of the family of stochastic languages. *Information and Control* **18**, 253–256 (1971).

36. M. Nasu and N. Honda, Fuzzy events realized by probabilistic automata. *Information and Control* **12**, 284–303 (1966).
37. M. Nasu and N. Honda, A context-free language which is not acceptable by a probabilistic automata. *Information and Control* **18**, 233–236 (1971).
38. P. H. Swain and K. S. Fu, Stochastic programmed grammars for syntactic pattern recognition. *Pattern Recognition* **4**, 83–100 (1972).
39. P. Suppes, Probabilistic grammars for natural languages. *Synthese* **22**, 95–116 (1970).
40. T. L. Booth and R. A. Thompson, Applying probability measures to abstract languages. *IEEE Trans. Computers* **C-22**, 442–450 (1973).

Chapter 6

Stochastic Languages for Syntactic Pattern Recognition

6.1 Stochastic Languages for Pattern Description

As was pointed out in Section 5.1, noisy patterns often cause ambiguity in their language representations. Also, in a particular class of patterns, some patterns may occur more often than others. In order to model the process more realistically, the use of stochastic languages for pattern description has been recently proposed[1–3].† In the case of an ambiguity, that is, when a string has two or more parses, we can select, for example, the most probable parse as the syntactic description of the string. The most probable parse is defined as the parse according to which the string will be generated with the highest probability. If the strings $x \in L(\bar{G}_s)$ representing patterns and their associated string probabilities $p(x)$ are known (or can be estimated), it would be possible to calculate or to estimate the production probabilities, so that the resulting stochastic grammar G_s will generate the strings with (approximately) the same probabilities. Furthermore, the assignment of probability to

---

† Since the approach was proposed very recently, it can only be demonstrated by a series of examples at this time. There is no doubt that many more studies have to be done before we can conclusively evaluate this approach.

each string generated by a grammar can also be used to control the "unwanted" strings. Very small values of probability could be assigned to the unwanted strings such that the resulting stochastic grammar would generate the wanted strings with high probability. Several illustrative examples are presented in this section.

   Example 6.1   An equilateral triangle and eight other distorted, or noisy, versions are shown in Fig. 6.1. The pattern primitives selected are given in Fig. 6.2. Each triangle is described by a string of length 3. Suppose that, from a priori knowledge or actual observations, the probabilities of the generation of these nine different triangles are known or can be estimated, say, from the relative frequencies of occurrences of these triangles. This information is listed in Table 6.1.

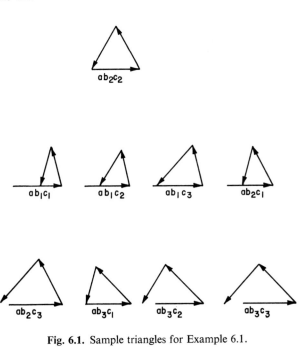

Fig. 6.1. Sample triangles for Example 6.1.

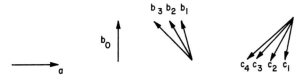

Fig. 6.2. Pattern primitives used in Examples 6.1 and 6.2.

TABLE 6.1

x	$p(x)$	x	$p(x)$
$x_1 = ab_1c_1$	1/36	$x_6 = ab_2c_3$	2/36
$x_2 = ab_1c_2$	2/36	$x_7 = ab_3c_1$	3/36
$x_3 = ab_1c_3$	3/36	$x_8 = ab_3c_2$	2/36
$x_4 = ab_2c_1$	1/36	$x_9 = ab_3c_3$	1/36
$x_5 = ab_2c_2$	21/36		

A stochastic finite-state grammar which will generate these strings is†
$G_1 = (V_{N1}, V_{T1}, P_1, S)$ where

$$V_{N1} = \{S, A_1, A_2, A_3, A_4\}, \quad V_{T1} = \{a, b_1, b_2, b_3, c_1, c_2, c_3\},$$

and P_1:

(1) $S \xrightarrow{p_1} aA_1$ (8) $A_3 \xrightarrow{p_8} c_1$

(2) $A_1 \xrightarrow{p_2} b_1A_2$ (9) $A_3 \xrightarrow{p_9} c_2$

(3) $A_1 \xrightarrow{p_3} b_2A_3$ (10) $A_3 \xrightarrow{p_{10}} c_3$

(4) $A_1 \xrightarrow{p_4} b_3A_4$ (11) $A_4 \xrightarrow{p_{11}} c_1$

(5) $A_2 \xrightarrow{p_5} c_1$ (12) $A_4 \xrightarrow{p_{12}} c_2$

(6) $A_2 \xrightarrow{p_6} c_2$ (13) $A_4 \xrightarrow{p_{13}} c_3$

(7) $A_2 \xrightarrow{p_7} c_3$

The normalization conditions are

$$p_1 = 1, \quad p_2 + p_3 + p_4 = 1, \quad p_5 + p_6 + p_7 = 1,$$
$$p_8 + p_9 + p_{10} = 1, \quad p_{11} + p_{12} + p_{13} = 1$$

In order to generate the string $x_1 = ab_1c_1$, productions (1), (2), and (5) must be applied; hence,

$$p(x_1) = p_1p_2p_5 = 1/36$$

Similarly,

$$p(x_2) = p_1p_2p_6 = 2/36, \qquad p(x_6) = p_1p_3p_{10} = 2/36$$
$$p(x_3) = p_1p_2p_7 = 3/36, \qquad p(x_7) = p_1p_4p_{11} = 3/36$$
$$p(x_4) = p_1p_3p_8 = 1/36, \qquad p(x_8) = p_1p_4p_{12} = 2/36$$
$$p(x_5) = p_1p_3p_9 = 21/36, \qquad p(x_9) = p_1p_4p_{13} = 1/36$$

† Refer to Section 7.5 for the inference of the grammar.

From $p_1 = 1$, $p_1 p_2 (p_5 + p_6 + p_7) = 1/36 + 2/36 + 3/36 = 1/6$, and $p_5 + p_6 + p_7 = 1$, we obtain $p_2 = 1/6$, $p_5 = 1/6$, $p_6 = 1/3$, $p_7 = 1/2$. Similarly, we obtain

$$p_3 = 2/3, \qquad p_8 = 1/24, \qquad p_9 = 21/24, \qquad p_{10} = 1/12$$
$$p_4 = 1/6, \qquad p_{11} = 1/2, \qquad p_{12} = 1/3, \qquad p_{13} = 1/6.$$

The grammar G_1 with the foregoing production probabilities will generate strings with associated probabilities as shown in Table 6.1.

Example 6.2 A right-angled triangle and eight other distorted or noisy versions are shown in Fig. 6.3. Based on the pattern primitives shown in Fig. 6.2 and the probability information listed in Table 6.2, the stochastic

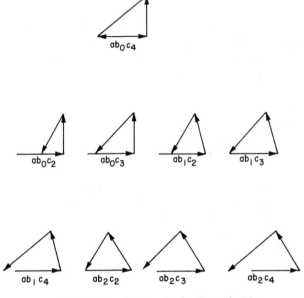

Fig. 6.3. Sample triangles for Example 6.2.

TABLE 6.2

x	$p(x)$	x	$p(x)$
ab_0c_2	1/36	ab_1c_4	1/36
ab_0c_3	2/36	ab_2c_2	2/36
ab_0c_4	21/36	ab_2c_3	3/36
ab_1c_2	1/36	ab_2c_4	1/36
ab_1c_3	4/36		

finite-state grammar that will generate these nine strings with associated probabilities is $G_2 = (V_{N2}, V_{T2}, P_2, S)$, where

$$V_{N2} = \{S, A_1, A_2, A_3, A_4\}, \quad V_{T2} = \{a, b_0, b_1, b_2, c_2, c_3, c_4\},$$

and P_2:

$$S \xrightarrow{\ 1\ } aA_1 \qquad A_2 \xrightarrow{\ 1/12\ } c_3 \qquad A_3 \xrightarrow{\ 1/6\ } c_4$$

$$A_1 \xrightarrow{\ 2/3\ } b_0 A_2 \qquad A_2 \xrightarrow{\ 21/24\ } c_4 \qquad A_4 \xrightarrow{\ 1/3\ } c_2$$

$$A_1 \xrightarrow{\ 1/6\ } b_1 A_3 \qquad A_3 \xrightarrow{\ 1/6\ } c_2 \qquad A_4 \xrightarrow{\ 1/2\ } c_3$$

$$A_1 \xrightarrow{\ 1/6\ } b_2 A_4 \qquad A_3 \xrightarrow{\ 2/3\ } c_3 \qquad A_4 \xrightarrow{\ 1/6\ } c_4$$

$$A_2 \xrightarrow{\ 1/24\ } c_2$$

It is noted that the strings ab_1c_2, ab_1c_3, ab_2c_2, and ab_2c_3 in Table 6.2 are also listed in Table 6.1. If we consider the patterns described by the strings in Table 6.1 as forming the pattern class ω_1 and the patterns described by the strings in Table 6.2 as forming the pattern class ω_2, then the four patterns ab_1c_2, ab_1c_3, ab_2c_2, and ab_2c_3 fall into both classes. For the nonstochastic case, the finite-state automata (or syntax analyzers) designed on the basis of $\bar{G}_1$ and $\bar{G}_2$ will not be able to discriminate these four strings; that is, they both accept the four strings. In such a circumstance, an idea similar to that used in the decision-theoretic approach can be applied. The probability information $p(x|\omega_i)$, $i = 1, 2$, about the strings belonging to each pattern class plays an important role in the recognition problem. For example, suppose that the pattern to be recognized is represented by the string ab_1c_2. With the assumption of equal a priori probabilities (i.e., $P(\omega_1) = P(\omega_2)$), the information $p(x|G_i) = p(x|\omega_i)$, $i = 1, 2$, can be used for the maximum-likelihood classification rule. In this case, since

$$p(ab_1c_2|G_1) = 2/36, \quad p(ab_1c_2|G_2) = 1/36, \quad \text{and} \quad p(ab_1\,c_2|G_1) > p(ab_1c_2|G_2)$$

the pattern represented by string ab_1c_2 should be classified as from the class ω_1.

Since the probability information $p(x|G_i)$ is used in the classification rule, the computation of $p(x|G_i)$ for any given string x at the time of classification becomes an important problem. In an extremely simple problem, such as that in Example 6.2, the information in Tables 6.1 and 6.2 (which can be easily derived from the two corresponding stochastic grammars) can be directly stored in the recognition system. A "table look-up" procedure will provide the necessary information for classification. However, in general, the number of strings generated by a grammar is often very large or infinite. One of the advantages of using a grammar to describe the structure of a language is that such a finite representation can be used instead of listing all the strings in the language. For finite-state languages, finite-state automata can be designed for the recognition purpose. For example, applying the procedure introduced in Section 5.3.1, we can construct the stochastic finite-state automaton S_{a1} which will accept strings generated by G_1 in Example 6.1.

$$S_{a1} = (\Sigma, Q, M, \pi_0, F)$$

where $\Sigma = \{a, b_1, b_2, b_3, c_1, c_2, c_3\}$, $Q = \{S, A_1, A_2, A_3, A_4, T, R\}$, $F = \{T\}$, $\pi_0 = [1, 0, 0, 0, 0, 0, 0]$, and

$$M(a) = \begin{bmatrix} 0 & 1 & 0 & 0 & 0 & 0 & 0 \\ 0 & 0 & 0 & 0 & 0 & 0 & 1 \\ 0 & 0 & 0 & 0 & 0 & 0 & 1 \\ 0 & 0 & 0 & 0 & 0 & 0 & 1 \\ 0 & 0 & 0 & 0 & 0 & 0 & 1 \\ 0 & 0 & 0 & 0 & 0 & 0 & 1 \\ 0 & 0 & 0 & 0 & 0 & 0 & 1 \end{bmatrix}, \quad M(b_1) = \begin{bmatrix} 0 & 0 & 0 & 0 & 0 & 0 & 1 \\ 0 & 0 & \frac{1}{6} & 0 & 0 & 0 & \frac{5}{6} \\ 0 & 0 & 0 & 0 & 0 & 0 & 1 \\ 0 & 0 & 0 & 0 & 0 & 0 & 1 \\ 0 & 0 & 0 & 0 & 0 & 0 & 1 \\ 0 & 0 & 0 & 0 & 0 & 0 & 1 \\ 0 & 0 & 0 & 0 & 0 & 0 & 1 \end{bmatrix}$$

$$M(b_2) = \begin{bmatrix} 0 & 0 & 0 & 0 & 0 & 0 & 1 \\ 0 & 0 & 0 & \frac{2}{3} & 0 & 0 & \frac{1}{3} \\ 0 & 0 & 0 & 0 & 0 & 0 & 1 \\ 0 & 0 & 0 & 0 & 0 & 0 & 1 \\ 0 & 0 & 0 & 0 & 0 & 1 & 1 \\ 0 & 0 & 0 & 0 & 0 & 0 & 1 \\ 0 & 0 & 0 & 0 & 0 & 0 & 1 \end{bmatrix}, \quad M(b_3) = \begin{bmatrix} 0 & 0 & 0 & 0 & 0 & 0 & 1 \\ 0 & 0 & 0 & 0 & \frac{1}{6} & 0 & \frac{5}{6} \\ 0 & 0 & 0 & 0 & 0 & 0 & 1 \\ 0 & 0 & 0 & 0 & 0 & 0 & 1 \\ 0 & 0 & 0 & 0 & 0 & 0 & 1 \\ 0 & 0 & 0 & 0 & 0 & 0 & 1 \\ 0 & 0 & 0 & 0 & 0 & 0 & 1 \end{bmatrix}$$

$$M(c_1) = \begin{bmatrix} 0 & 0 & 0 & 0 & 0 & 0 & 1 \\ 0 & 0 & 0 & 0 & 0 & 0 & 1 \\ 0 & 0 & 0 & 0 & 0 & \frac{1}{6} & \frac{5}{6} \\ 0 & 0 & 0 & 0 & 0 & \frac{1}{24} & \frac{23}{24} \\ 0 & 0 & 0 & 0 & 0 & \frac{1}{2} & \frac{1}{2} \\ 0 & 0 & 0 & 0 & 0 & 0 & 1 \\ 0 & 0 & 0 & 0 & 0 & 0 & 1 \end{bmatrix}, \quad M(c_2) = \begin{bmatrix} 0 & 0 & 0 & 0 & 0 & 0 & 1 \\ 0 & 0 & 0 & 0 & 0 & 0 & 1 \\ 0 & 0 & 0 & 0 & 0 & \frac{1}{3} & \frac{2}{3} \\ 0 & 0 & 0 & 0 & 0 & \frac{21}{24} & \frac{3}{24} \\ 0 & 0 & 0 & 0 & 0 & \frac{1}{3} & \frac{2}{3} \\ 0 & 0 & 0 & 0 & 0 & 0 & 1 \\ 0 & 0 & 0 & 0 & 0 & 0 & 1 \end{bmatrix}$$

$$M(c_3) = \begin{bmatrix} 0 & 0 & 0 & 0 & 0 & 0 & 1 \\ 0 & 0 & 0 & 0 & 0 & 0 & 1 \\ 0 & 0 & 0 & 0 & 0 & \frac{1}{2} & \frac{1}{2} \\ 0 & 0 & 0 & 0 & 0 & \frac{1}{12} & \frac{11}{12} \\ 0 & 0 & 0 & 0 & 0 & \frac{1}{6} & \frac{5}{6} \\ 0 & 0 & 0 & 0 & 0 & 0 & 1 \\ 0 & 0 & 0 & 0 & 0 & 0 & 1 \end{bmatrix}, \quad \pi_F = \begin{bmatrix} 0 \\ 0 \\ 0 \\ 0 \\ 0 \\ 1 \\ 0 \end{bmatrix}$$

It is easy to verify that

$$p(ab_1c_2 \,|\, G_1) = \pi_0 M(ab_1c_2)\pi_F = \pi_0 M(a)M(b_1)M(c_2)\pi_F = 2/36$$

Similarly,

$$p(ab_1c_1 | G_1) = \pi_0 M(ab_1c_1)\pi_F = 1/36$$
$$p(ab_1c_3 | G_1) = \pi_0 M(ab_1c_3)\pi_F = 3/36$$
$$p(ab_2 c_1 | G_1) = \pi_0 M(ab_2 c_1)\pi_F = 1/36$$
$$p(ab_2 c_2 | G_1) = \pi_0 M(ab_2 c_2)\pi_F = 21/36$$
$$p(ab_2 c_3 | G_1) = \pi_0 M(ab_2 c_3)\pi_F = 2/36$$
$$p(ab_3 c_1 | G_1) = \pi_0 M(ab_3 c_1)\pi_F = 3/36$$
$$p(ab_3 c_2 | G_1) = \pi_0 M(ab_3 c_2)\pi_F = 2/36$$
$$p(ab_3 c_3 | G_1) = \pi_0 M(ab_3 c_3)\pi_F = 1/36$$

For patterns described by context-free languages, syntax analysis is required for the input string (a pattern) with respect to each grammar G_i. After all the parses are obtained, the probabilities of x generated by each grammar, $p(x | G_i)$, can be calculated. (If the string x is rejected with respect to grammar G_j, then $p(x | G_j) = 0$.) Then, the maximum-likelihood classification rule is applied, that is, the pattern represented by string x will be assigned to class ω_k if

$$p(x | G_k) = \underset{G_i}{\text{Max}} \{ p(x | G_i) \} \tag{6.1}$$

The parse associated with G_k will provide the syntactic description of x. A block diagram of such a recognition scheme is shown in Fig. 6.4. More

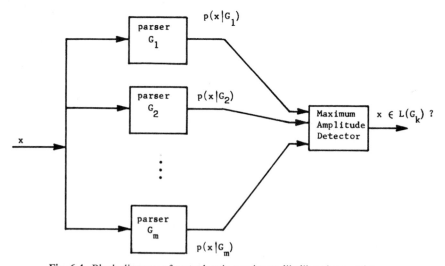

Fig. 6.4. Block diagram of a stochastic maximum-likelihood recognizer.

generally, classification rules can also be formulated with known a priori probabilities and/or given loss functions (e.g., Bayes's classification rule).

Example 6.3   A grammar is developed to describe "noisy squares," for example, the squares drawn on a cathode-ray tube (CRT) input device or a RAND tablet [4]. The following assumptions are made to keep the grammar simple but interesting.

(1)   The squares were to have their side length geometrically distributed with mean length $1/\alpha$ ($0 < \alpha < 1$).

(2)   The squares were to be drawn horizontally/vertically and the operator was assumed to be skillful enough so that the "noise" could be considered primarily the quantization error resulting from input through a digital or finite-state device.

(3)   The "target aiming" effect, which could come into play as a square is about to be closed and would assist in ensuring closure, was ignored.

It should be noted that the simplifying assumptions could be relaxed and the grammar made arbitrarily complex. However, the example as defined serves to illustrate that

(1)   a relatively complex context-sensitive language can be generated by a context-free programmed grammar;

(2)   a noisy physical situation can be modeled by a stochastic context-free programmed grammar;

(3)   a general program for simulating the operation of programmed grammars can aid in the interactive development of grammars intended to model physical situations.

A generator program for languages generated by stochastic context-free programmed grammars was written for a PDP-9 digital computer with a Model 339 Graphical Display. The program generates strings stochastically using any specified stochastic context-free programmed grammar as input "data," and displays a pictorial interpretation of the generation process. The generation and display capabilities permit the user to observe the grammar at work, to interactively improve the grammar, and to qualitatively assess its stochastic properties.

For this example, a language with tails was used.† A typical noisy square that might be generated is shown in Fig. 6.5, and the stochastic context-free

---

† Let L_1 be a language over the terminals V_T and L_2 be a language over $V_T \cup \{s, t\}$ where $s, t \notin V_T$ such that $L_2 = \{xst^l \mid x \in L_1$ and l depends on $x\}$. Then L_2 is said to be of the form L_1 with tails. The x portion is referred to as the body of the string, s as the tail delimiter, and t^l as the tail. In general, the tail is found to be a convenient place to perform any bookkeeping functions (as a counter in this example) necessary for the generation of the language.

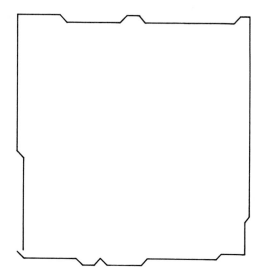

Fig. 6.5. A typical noisy square. Scale: $\frac{1}{16}$ inch per unit length; $\alpha = 0.05$, $\beta = 0.05$.

programmed grammar appears in Table 6.3. "Reasonable" values were selected for the branch probabilities in the grammar. If a hand-drawn "training set" were available for analysis in terms of the grammatical model, these probabilities could be more closely adjusted to the actual physical situation.

Example 6.4   This example illustrates another use of stochastic context-free programmed grammars for pattern description. The following three stochastic context-free programmed grammars are developed for the description of

TABLE 6.3

Stochastic Grammar Describing Noisy Squares[a]

$$G_{sp} = (V_N, V_T, P, S, J)$$

where

$$V_N = \{S, F, X, Q, A, M, W, C, E, R, T, B, V\}, \quad V_T = \{-, 0, +, *, \$\},$$
$$J = \{1, 2, \ldots, 37\}$$

and

0:   unit segment in direction of current side
+ :   unit segment with slope $+1$ relative to current side
− :   unit segment
∗ :   corner (change direction by $+90°$)
\$:   tail delimiter (final string only)

TABLE 6.3   *Continued*

Label	Core	Success branches and branch probabilities[b]	Failure branches and branch probabilities
1	$S \to FXVQ$	$\{2, 4, 7\}, \{1 - 2\alpha, \alpha, \alpha\}$	$\{\varnothing\}, \{1\}$
2	$F \to 0FA$	$\{3\}, \{1\}$	$\{\varnothing\}, \{1\}$
3	$X \to XXX$	$\{10\}, \{1\}$	$\{\varnothing\}, \{1\}$
4	$F \to + FA$	$\{5\}, \{1\}$	$\{\varnothing\}, \{1\}$
5	$M \to XXX$	$\{10\}, \{1\}$	$\{6\}, \{1\}$
6	$X \to WXX$	$\{10\}, \{1\}$	$\{\varnothing\}, \{1\}$
7	$F \to - FA$	$\{8\}, \{1\}$	$\{\varnothing\}, \{1\}$
8	$W \to XXX$	$\{10\}, \{1\}$	$\{9\}, \{1\}$
9	$X \to MXX$	$\{10\}, \{1\}$	$\{\varnothing\}, \{1\}$
10	$W \to W$	$\{2, 4, 7, 12\}, (p_1, p_2, p_3, \beta\}$	$\{11\}, \{1\}$
11	$M \to M$	$\{2, 4, 7, 12\}, \{p_1, p_3, p_2, \alpha\}$	$\{2, 4, 7, 12\}, \{p_4, p_3, p_3, \beta\}$
12	$F \to * F$	$\{13\}, \{1\}$	$\{\varnothing\}, \{1\}$
13	$Q \to R$	$\{16\}, \{1\}$	$\{14\}, \{1\}$
14	$R \to S$	$\{16\}, \{1\}$	$\{15\}, \{1\}$
15	$S \to T$	$\{16\}, \{1\}$	$\{36\}, \{1\}$
16	$B \to A$	$\{16\}, \{1\}$	$\{17\}, \{1\}$
17	$A \to B$	$\{18\}, \{1\}$	$\{20\}, \{1\}$
18	$E \to C$	$\{17\}, \{1\}$	$\{19\}, \{1\}$
19	$V \to CV$	$\{17\}, \{1\}$	$\{\varnothing\}, \{1\}$
20	$W \to X$	$\{21\}, \{1\}$	$\{22\}, \{1\}$
21	$C \to E$	$\{20\}, \{1\}$	$\{\varnothing\}, \{1\}$
22	$M \to X$	$\{23\}, \{1\}$	$\{25\ 26\ 29\}, \{1 - 2\alpha, \alpha, \alpha\}$
23	$E \to C$	$\{22\}, \{1\}$	$\{24\}, \{1\}$
24	$V \to CV$	$\{22\}, \{1\}$	$\{\varnothing\}, \{1\}$
25	$F \to 0F$	$\{32\}, \{1\}$	$\{\varnothing\}, \{1\}$
26	$F \to +F$	$\{27\}, \{1\}$	$\{\varnothing\}, \{1\}$
27	$M \to X$	$\{32\}, \{1\}$	$\{28\}, \{1\}$
28	$X \to W$	$\{32\}, \{1\}$	$\{\varnothing\}, \{1\}$
29	$F \to -F$	$\{30\}, \{1\}$	$\{\varnothing\}, \{1\}$
30	$W \to X$	$\{32\}, \{1\}$	$\{31\}, \{1\}$
31	$X \to M$	$\{32\}, \{1\}$	$\{\varnothing\}, \{1\}$
32	$C \to E$	$\{33\}, \{1\}$	$\{\varnothing\}, \{1\}$
33	$C \to C$	$\{34\}, \{1\}$	$\{12\}, \{1\}$
34	$W \to W$	$\{25, 26, 29\}, \{1 - \alpha - \alpha^2, \alpha^2, \alpha\}$	$\{35\}, \{1\}$
35	$M \to M$	$\{25, 26, 29\}, \{1 - \alpha - \alpha^2, \alpha, \alpha^2\}$	$\{25, 26, 29\}, \{1 - 2\alpha, \alpha, \alpha\}$
36	$T \to \$$	$\{37\}, \{1\}$	$\{\varnothing\}, \{1\}$
37	$F \to F$	$\{\varnothing\}, \{1\}$	$\{\varnothing\}, \{1\}$

[a] The noisy square shown in Fig. 6.5 is the pictorial representation of the string:

$-000000000 - 00 + -000000 + 00000000000 + 0000*$

$00000 - 0000000000000000000000000000000000000*$

$00 + 00000000000000 - 00 + 000000000 - 0000000*$

$000000000000000000000 + 000000000000000*\$$ (tail omitted)

[b] Here $p_1 = (1-\alpha-\alpha^2)(1-\beta)$, $p_2 = \alpha^2(1-\beta)$, $p_3 = \alpha(1-\beta)$, $p_4 = (1-2\alpha)(1-\beta)$.

median, submedian, and acrocentric chromosome patterns, respectively (see Fig. A.2). The grammars are heuristically constructed to avoid ambiguity and the possible generation of unwanted patterns, such as those shown in Fig. 6.6. (Refer to Appendix A for information on using context-free grammars for the description of chromosome patterns.)

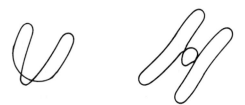

Fig. 6.6. Unwanted chromosome patterns.

The probability distribution associated with each go-to field of each of the three chromosome grammars determines the probability distribution of chromosome patterns defined in each of the three chromosome pattern classes. Since we have constructed the chromosome grammars such that some median type chromosomes are considered as the distorted and/or noisy versions of acrocentric type chromosomes, and median type and acrocentric type chromosomes are considered to be the distorted and/or noisy versions of submedian type chromosomes, then the set of probabilities $(p_{1m}, \ldots, p_{4m}, p_{1s}, \ldots, p_{7s}, p_{1a}, \ldots, p_{5a})$ is chosen such that (i) a median type chromosome is more likely to be generated by G_m than by G_s or G_a, (ii) a submedian type chromosome is more likely to be generated by G_s than by G_m or G_a, and (iii) an acrocentric type chromosome is more likely to be generated by G_a than by G_m or G_s.

Median Chromosomes

$$G_m = (V_{Tm}, V_{Nm}, J_m, P_m, S)$$

where

$$V_{Tm} = \{a, b, c, d\}$$

where $a, b, c,$ and d are primitives defined in Fig. A.1.

$$V_{Nm} = \{S, A, B, C, D\}, \qquad J_m = \{1, \ldots, 25\}$$

and P_m is defined as follows:

Label	Core	U	$P(U)$
1	$S \rightarrow ABC$	2	1
2	$A \rightarrow cA$	3	1
3	$B \rightarrow BcB$	4, 10	$p_{1m}, 1 - p_{1m}$
4	$C \rightarrow Cb$	5	1
5	$A \rightarrow bA$	6	1
6	$B \rightarrow Db$	7	1
7	$B \rightarrow bD$	8	1
8	$D \rightarrow B$	9	1
9	$D \rightarrow B$	4, 10	$p_{2m}, 1 - p_{2m}$
10	$A \rightarrow aA$	11	1
11	$C \rightarrow Ca$	12	1
12	$B \rightarrow Da$	13	1
13	$B \rightarrow aD$	14	1
14	$D \rightarrow B$	15	1
15	$D \rightarrow B$	16, 22	$p_{3m}, 1 - p_{3m}$
16	$A \rightarrow A$	17	1
17	$C \rightarrow Cb$	18	1
18	$B \rightarrow Db$	19	1
19	$B \rightarrow bD$	20	1
20	$D \rightarrow B$	21	1
21	$D \rightarrow B$	22, 16	$p_{4m}, 1 - p_{4m}$
22	$A \rightarrow b$	23	1
23	$C \rightarrow b$	24	1
24	$B \rightarrow db$	25	1
25	$B \rightarrow bd$	$\varnothing$	1

Submedian Chromosomes

$$G_s = (V_{Ts}, V_{Ns}, J_s, P_s, S)$$

where $V_{Ts} = V_{Tm}$, $V_{Ns} = \{S, A, B\}$, $J_s = \{1, \ldots, 20\}$, P_s is defined as follows:

Label	Core	U	$P(U)$
1	$S \rightarrow AB$	2	1
2	$A \rightarrow cA$	3, 5	$p_{1s}, 1 - p_{1s}$
3	$A \rightarrow bA$	4	1
4	$B \rightarrow Bb$	3, 5	$p_{2s}, 1 - p_{2s}$
5	$A \rightarrow aA$	6	1
6	$B \rightarrow Ba$	7	1
7	$A \rightarrow bA$	8	1
8	$B \rightarrow Bb$	7, 9	$p_{3s}, 1 - p_{3s}$
9	$A \rightarrow dA$	10	1

Label	Core	U	$P(U)$
10	$A \to Bd$	11	1
11	$A \to bA$	12	1
12	$B \to Bb$	11, 13, 17	$p_{4s}, p_{5s}, 1 - p_{4s} - p_{5s}$
13	$A \to aA$	14	1
14	$B \to Ba$	15, 19	$p_{6s}, 1 - p_{6s}$
15	$A \to bA$	16	1
16	$B \to Bb$	15, 19	$p_{7s}, 1 - p_{7s}$
17	$A \to a$	18	1
18	$B \to ca$	$\varnothing$	1
19	$A \to b$	20	1
20	$B \to cb$	$\varnothing$	1

Acrocentric Chromosomes

$$G_a = (V_{Ta}, V_{Na}, J_a, P_a, S)$$

where $V_{Ta} = V_{Tm}$, $V_{Na} = \{S, A, B\}$, $J_a = \{1, \ldots, 18\}$, and P_a is defined as follows:

Label	Core	U	$P(U)$
1	$S \to AB$	2	1
2	$A \to cA$	3	1
3	$A \to aA$	4	1
4	$B \to Ba$	5	1
5	$A \to bA$	6	1
6	$B \to Bb$	5, 7	$p_{1a}, 1 - p_{1a}$
7	$A \to dA$	8	1
8	$B \to Bd$	9	1
9	$A \to bA$	10	1
10	$B \to Bb$	9, 11, 15	$p_{2a}, p_{3a}, 1 - p_{2a} - p_{3a}$
11	$A \to aA$	12	1
12	$B \to Ba$	13, 17	$p_{4a}, 1 - p_{4a}$
13	$A \to bA$	14	1
14	$B \to Bb$	13, 17	$p_{5a}, 1 - p_{5a}$
15	$A \to a$	16	1
16	$B \to ca$	$\varnothing$	1
17	$A \to b$	18	1
18	$B \to cb$	$\varnothing$	1

6.2 Estimation of Production Probabilities

This section describes procedures that are used to estimate or infer production probabilities from a given characteristic (nonstochastic) grammar and sample information. The sample information usually consists of a set of distinct sample strings and their associated string probabilities. The string probabilities can be estimated or subjectively assigned. Let the sample information be

$$S_t = \{(x_1, f_1), \ldots, (x_t, f_t)\} \qquad (6.2)$$

where f_k is the estimated or subjective probability† of x_k. Two types of estimation procedures are discussed [5, 6].

6.2.1   Single-Class Estimation

In this case, the information about the grammar generating each sample string is known. The problem is essentially a single-class (or one-grammar) problem, since only the sample strings known to be generated by the same grammar are used in the estimation. Let the characteristic (context-free) grammar be

$$G = (V_N, V_T, P, S)$$

and p_{ij} be the production probability associated with the production

$$A_i \to \eta_j, \qquad A_i \in V_N, \qquad \eta_j \in (V_N \cup V_T)^+$$

Since a stochastic context-free grammar can be modeled by a multitype Galton–Watson branching process, the theory of statistical inference of Markov processes can be applied to infer the production probabilities [7]. Let the inferred or estimated production probabilities be $\hat{p}_{ij}$ and $\bar{S}_t = \{x_1, \ldots, x_t\}$, $\bar{S}_t \subseteq L(G)$. Then the maximum-likelihood estimate for p_{ij} is

$$\hat{p}_{ij} = \frac{n_{ij}}{\sum_j n_{ij}} \qquad (6.3)$$

where n_{ij} is the expected number of times that the production $A_i \to \eta_j$ in G is used in parsing all the sample strings in $\bar{S}_t$.

$$n_{ij} = \sum_{x_k \in S_t} f_k N_{ij}(x_k) \qquad (6.4)$$

$N_{ij}(x_k)$ is the number of times that the production $A_i \to \eta_j$ is used in parsing string x_k. Substituting (6.4) into (6.3), we obtain

$$\hat{p}_{ij} = \frac{\sum_{x_k \in \bar{S}_t} f_k N_{ij}(x_k)}{\sum_j \sum_{x_k \in \bar{S}_t} f_k N_{ij}(x_k)} \qquad (6.5)$$

A block diagram of the production probability inference system is shown in Fig. 6.7.

---

† In the case of controlling unwanted strings, subjective probability assignment is often used.

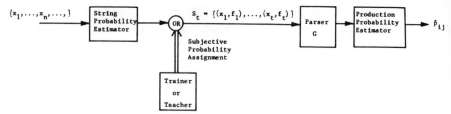

Fig. 6.7. Block diagram of a single-class production probability inference system.

It can be shown that, under the following conditions, $\hat{p}_{ij}$ approaches p_{ij} as the number of sample strings t approaches infinity [5].

(i)   The sample set $\bar{S}_t$ approaches $L(G)$ as t increases.

(ii)   The estimated or subjective probability of x_k, f_k, approaches the true probability $p(x_k)$ as t increases.

Computation of all p_{ij}'s is made from a knowledge of the n_{ij}'s. If at a later time a better estimate is desired through the consideration of more sample strings, only the n_{ij}'s need to be saved from the previous estimate. All previously considered strings need not be parsed again. Let $n_{ij}(t)$ be calculated by (6.4) from S_t. If a new string x_{t+1} is added to $\bar{S}_t$ to form $\bar{S}_{t+1}$, then $n_{ij}(t + 1)$ can be updated by the following recursive equation:

$$n_{ij}(t + 1) = n_{ij}(t) + N_{ij}(x_{t+1})f_{t+1} \qquad (6.6)$$

The following example illustrates an application of the single-class estimation procedure.

Example 6.5   Let the stochastic source grammar be $G_s = (V_N, V_T, P_s, S)$, where $V_N = \{A_1, A_2\}$, $V_T = \{a, b, c, d\}$, $S = \{A_1\}$, and P_s:

$$A_1 \xrightarrow{0.2} bA_2 A_1 \qquad A_2 \xrightarrow{0.9} a$$

$$A_1 \xrightarrow{0.8} dA_2 \qquad A_2 \xrightarrow{0.1} cA_2$$

The grammar is consistent and unambiguous.

(1)   Three string sample sizes, 10, 100, and 200, are generated by the source grammar. The generated strings with their number of occurrences are listed in Table 6.4. Only about 11 different strings are produced in a sample size of 200 strings, although the set of strings that the grammar can generate is infinite. Frequency ratios are used as the estimated string probabilities. The grammar whose production probabilities are to be estimated is $G_s' =$

TABLE 6.4

<small>SAMPLE STRINGS FOR EXAMPLE 6.5</small>

	No. of Occurrences for sample size			True string
Sample string x_k	10	100	200	Probability $p(x_k)$
bada	1	9	27	0.12960
da	8	77	146	0.72000
badca	1	2	4	0.01296
dca		6	11	0.07200
bccabadca		1	1	0.00002
bcada		2	2	0.01296
bcababada		1	1	0.00042
babada		2	5	0.02333
bcabcada			1	0.00023
dcca			1	0.00720
bcabada			1	0.00233

(V_N, V_T, P_s', S), where $V_N = \{A_1, A_2\}$, $V_T = \{a, b, c, d\}$, $S = \{A_1\}$, and P_s':

$$A_1 \xrightarrow{p_{11}} bA_2A_1 \qquad A_2 \xrightarrow{p_{21}} a$$

$$A_1 \xrightarrow{p_{12}} dA_2 \qquad A_2 \xrightarrow{p_{22}} cA_2$$

In the actual calculation of $\hat{p}_{ij}$ based on (6.5), the number of occurrences of string x_k for a given sample size is used for f_k. The estimation results are given in the following.

Estimated production probability	Sample size			True production probability
	10	100	200	
$\hat{p}_{11}$	0.16667	0.18033	0.20635	$p_{11} = 0.2$
$\hat{p}_{12}$	0.83333	0.81967	0.79365	$p_{12} = 0.8$
Confidence for $\hat{p}_{1j}$	78%	60%	80%	
$\hat{p}_{21}$	0.92308	0.89706	0.90648	$p_{21} = 0.9$
$\hat{p}_{22}$	0.07692	0.10294	0.09352	$p_{22} = 0.1$
Confidence for $\hat{p}_{2j}$	79%	90%	72%	

(2) If subjective probabilities can be assigned (by a teacher, say) to the strings, we could use only the 11 different strings produced from the 200 samples. In this case, let the subjective probabilities of the 11 strings be the

same as their true generation probabilities, that is, $f_k = p(x_k)$ in (6.5). The estimation results are given in the following.

Estimated production probability	Sample size			True production probability
	3	8	11	
$\hat{p}_{11}$	0.142	0.173	0.175	0.2
$\hat{p}_{12}$	0.858	0.827	0.825	0.8
$\hat{p}_{21}$	0.987	0.923	0.911	0.9
$\hat{p}_{22}$	0.013	0.077	0.089	0.1

It can be easily checked that G_s' is also consistent (Section 5.2.2). For this simple example grammar, it is possible to solve for the production probabilities from the estimated or subjective probabilities of strings. (Refer to Example 6.1.) However, this method of using the solution of nonlinear simultaneous equations is not straightforward when the grammar contains many productions or S_t contains a large number of sample strings.

6.2.2   Multiclass Estimation

If the sample strings are only known to be generated by context-free grammars $G_1, G_2, \ldots, G_m$, the following multiclass estimation procedure may be used. Let

$$G_q = (V_{Nq}, V_{Tq}, P_q, S_q), \qquad q = 1, \ldots, m \qquad (6.7)$$

and the set of sample strings $\bar{S}_t \subseteq L$ where $L = \bigcup_q L(G_q)$. A block diagram of the production probability inference system is shown in Fig. 6.8. A trainer or teacher is required to provide the information of $P(G_q|x_k)$, the probability that the given x_k belongs to class ω_q (or is generated by grammar G_q), $k = 1, \ldots, t, q = 1, \ldots, m$.

$$\sum_{q=1}^{m} P(G_q|x_k) = 1 \qquad (6.8)$$

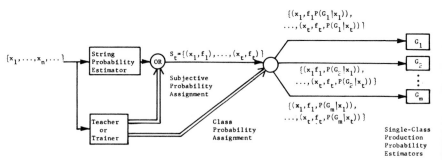

Fig. 6.8. Block diagram of a multiclass production probability inference system.

Then for grammar G_q,

$$n_{qij} = \sum_{x_k \in S_t} f_k P(G_q | x_k) N_{qij}(x_k) \tag{6.9}$$

where n_{qij} is the expected number of times that the production $A_i \rightarrow \eta_j$ in P_q is used in parsing all the sample strings, and $N_{qij}(x_k)$ is the number of times that the production $A_i \rightarrow \eta_j$ in P_q is used in the parse of string x_k. The maximum-likelihood estimate of p_{qij}, the production probability associated with $A_i \rightarrow \eta_j$ in P_q, is

$$\hat{p}_{qij} = \frac{n_{qij}}{\sum_j n_{qij}} \tag{6.10}$$

As the number of sample strings t increases, the following assumptions are made.

(1)   The set $\bar{S}_t$ approaches L;
(2)   the trainer can assign the proper $P(G_q | x_k)$ for each string x_k and grammar G_q; and
(3)   the estimated or subjective probability of x_k, f_k, approaches the true probability $p(x_k)$, where

$$p(x_k) = \sum_{q=1}^{m} p(x_k | G_q) P(G_q) \tag{6.11}$$

It can be shown that, under these assumptions, $\hat{p}_{qij}$ approaches p_{qij} as the number of sample strings t approaches infinity [5, 6].

It is noted that n_{ij} in (6.3) or n_{qij} in (6.10) follows a multinomial distribution with probability p_{ij} or p_{qij}, respectively. The chi-square test for goodness of fit can be applied to obtain the confidence region of the estimate [8]. Refer to (6.3); then

$$\chi^2 = \sum_{j=1}^{M} \frac{(\sum_j n_{ij})(\hat{p}_{ij} - p_{ij})^2}{p_{ij}} \tag{6.12}$$

Equation (6.12) can be used in a chi-square distribution with $(M - 1)$ degrees of freedom. M is the number of productions with A_i at the left side. The following example illustrates an application of the multiclass estimation procedure.

Example 6.6   The three stochastic source grammars are†

$$G_{s1} = (V_N, V_T, P_{s1}, S), \qquad G_{s2} = (V_N, V_T, P_{s2}, S),$$
$$G_{s3} = (V_N, V_T, P_{s3}, S)$$

---

† It should be noted that the three stochastic grammars have the same characteristic grammar, which is formed by the union of three individual characteristic grammars, but with different production probability assignments.

where $V_N = \{S, A, B\}$, $V_T = \{a, b\}$, and

$$P_{s1} = (P, D_1), \qquad P_{s2} = (P, D_2), \qquad P_{s3} = (P, D_3)$$

P	D_1	D_2	D_3
$S \to aB$	0.5	0.8	0.2
$S \to bA$	0.5	0.2	0.8
$A \to a$	0.9	0.7	0.7
$A \to aS$	0.05	0.2	0.05
$A \to bAA$	0.05	0.1	0.25
$B \to b$	0.9	0.7	0.7
$B \to bS$	0.05	0.05	0.2
$B \to aBB$	0.05	0.25	0.1

All three grammars are consistent and unambiguous.

TABLE 6.5

Stochastic-Generated Sample Strings for Multiclass Production Probability Inferences

		No. of occurrences of x_k for sample size			True string
	Strings (x_k)	10	100	200	probability $p(x_k)$
1	abab	1	2	5	1.346×10^{-2}
2	ba	5	41	79	3.790×10^{-1}
3	aabb	1	8	14	4.148×10^{-2}
4	ab	2	36	74	4.210×10^{-1}
5	bbabaa	1	2	2	4.091×10^{-3}
6	abbbaa		1	1	1.154×10^{-3}
7	bbaa		5	9	3.266×10^{-2}
8	abaabb		1	1	1.507×10^{-3}
9	baab		1	1	1.346×10^{-2}
10	bbaaba		1	1	1.154×10^{-3}
11	bbbbaaaaba		1	1	2.509×10^{-5}
12	bbbaabaa		1	1	6.351×10^{-4}
13	babbaa			1	1.154×10^{-3}
14	aaabbb			1	5.737×10^{-3}
15	aabaabbabb			1	3.691×10^{-5}
16	baaaabbb			1	2.226×10^{-4}
17	abba			1	1.178×10^{-2}
18	aababb			2	5.737×10^{-3}
19	aabbaaaabbbabb			1	2.537×10^{-7}
20	bbaababa			1	4.329×10^{-5}
21	bbbababababaabaa			1	7.391×10^{-7}
22	aabbba			1	8.607×10^{-4}

Let $P(G_{s1}) = P(G_{s2}) = P(G_{s3}) = 1/3$. Sample strings generated for sample sizes 10, 100, and 200 are listed in Table 6.5. The subjective probabilities of sample strings are assigned according to $p(x_k)$, that is, $f_k = p(x_k)$.

$$p(x_k) = \sum_{q=1}^{3} p(x_k|G_q)P(G_q) \quad \text{and} \quad P(G_q|x_k) = \frac{p(x_k|G_q)P(G_q)}{p(x_k)}$$

The estimation results for D_1, D_2, and D_3 are given in Table 6.6.

TABLE 6.6

	Class 1			Class 2			Class 3		
	True prob.	Sample size		True prob.	Sample size		True prob.	Sample size	
		10	100	200		10	100	200		10	100	200

I. Frequency ratios (number of occurrences) of strings used as the estimates of string probabilities

$S \to aB$	0.5	0.411	0.466	0.483	0.8	0.806	0.802	0.817	0.2	0.130	0.154	0.164
$S \to bA$	0.5	0.589	0.534	0.517	0.2	0.194	0.198	0.183	0.8	0.870	0.846	0.836
$A \to a$	0.9	0.937	0.900	0.911	0.7	0.875	0.797	0.779	0.7	0.704	0.689	0.685
$A \to aS$	0.05	0.00	0.020	0.020	0.2	0.00	0.078	0.120	0.05	0.00	0.035	0.057
$A \to bAA$	0.05	0.063	0.080	0.069	0.1	0.125	0.125	0.101	0.25	0.296	0.276	0.258
$B \to b$	0.9	0.749	0.880	0.867	0.7	0.692	0.775	0.698	0.7	0.704	0.725	0.700
$B \to bS$	0.05	0.158	0.044	0.056	0.05	0.127	0.051	0.064	0.2	0.189	0.186	0.178
$B \to aBB$	0.05	0.093	0.076	0.077	0.25	0.181	0.174	0.238	0.1	0.107	0.089	0.122

II. Subjective probabilities assigned according to $p(x_k)$

$S \to aB$	0.5	0.522	0.511	0.511	0.8	0.833	0.808	0.810	0.2	0.218	0.198	0.212
$S \to bA$	0.5	0.478	0.489	0.489	0.2	0.167	0.192	0.190	0.8	0.782	0.802	0.788
$A \to a$	0.9	0.996	0.934	0.934	0.7	0.990	0.824	0.824	0.7	0.947	0.840	0.838
$A \to aS$	0.05	0.00	0.022	0.026	0.2	0.00	0.115	0.116	0.05	0.00	0.010	0.014
$A \to bAA$	0.05	0.004	0.044	0.044	0.1	0.010	0.061	0.060	0.25	0.053	0.150	0.148
$B \to b$	0.9	0.941	0.940	0.914	0.7	0.866	0.862	0.816	0.7	0.915	0.895	0.786
$B \to bS$	0.05	0.021	0.023	0.042	0.005	0.025	0.028	0.030	0.2	0.031	0.052	0.150
$B \to aBB$	0.05	0.038	0.037	0.044	0.25	0.109	0.110	0.154	0.1	0.054	0.053	0.064

6.3   Examples of Stochastic Syntax Analysis

This section presents illustrative examples for stochastic syntax analysis (Sections 5.5 and 5.6).

Example 6.7   Two stochastic context-free grammars for describing abnormal (dicentric) and normal chromosome patterns, respectively, are

$$G_i = (V_N, V_T, P_{si}, S), \qquad i = 1, 2$$

where $V_N = \{N_1, N_2, \ldots, N_{23}\}$, $V_T = \{a, b, e, f, g, h, k, m, z, p, q\}$, $S = \{N_1\}$, and

$$P_{s1} = (P, D_1), \qquad P_{s2} = (P, D_2)$$

P	D_1	D_2	P	D_1	D_2
$N_1 \to N_2N_3N_4N_3$	0.656	0.552	$N_5 \to N_{16}$	0.5	0.970
$N_1 \to N_2N_3N_3$	0.031	0.431	$N_5 \to N_{17}$	0.5	0.030
$N_1 \to N_2N_3N_4N_4N_3$	0.312	0.017	$N_5 \to N_{18}$	1.2×10^{-7}	2.9×10^{-7}
$N_4 \to N_5N_6$	1.0	1.0	$N_{16} \to g$	1.0	1.0
$N_3 \to N_7N_6$	1.0	1.0	$N_{17} \to h$	1.0	1.0
$N_2 \to fN_6$	1.0	1.0	$N_{18} \to k$	1.0	1.0
$N_8 \to m$	0.656	0.233	$N_7 \to N_{19}$	0.047	0.767
$N_8 \to z$	0.344	0.767	$N_7 \to N_{20}$	0.953	0.233
$N_6 \to N_9$	0.482	0.832	$N_{19} \to qN_{21}q$	0.166	0.011
$N_6 \to N_{10}$	0.219	0.014	$N_{19} \to qN_{22}q$	0.166	0.011
$N_6 \to N_{11}$	0.150	0.005	$N_{19} \to qN_{23}q$	0.003	0.337
$N_6 \to N_{12}$	0.150	0.150	$N_{19} \to qN_{24}q$	0.666	0.640
$N_9 \to bN_{10}b$	0.391	0.883	$N_{24} \to bN_{23}b$	0.656	0.948
$N_9 \to bN_{11}b$	0.0086	0.011	$N_{24} \to bN_{24}b$	1.5×10^{-7}	1.7×10^{-7}
$N_9 \to bN_{12}b$	0.190	0.500	$N_{24} \to bN_{21}b$	0.016	0.017
$N_9 \to bN_9b$	0.431	0.006	$N_{24} \to bN_{22}b$	0.328	0.034
$N_{11} \to N_{10}b$	0.999	0.999	$N_{22} \to N_{23}b$	0.774	0.269
$N_{11} \to N_{11}b$	2.3×10^{-7}	3.3×10^{-6}	$N_{22} \to N_{22}b$	0.226	0.730
$N_{12} \to bN_{10}$	0.955	0.518	$N_{20} \to pN_{21}p$	0.492	0.038
$N_{12} \to bN_{12}$	0.045	0.482	$N_{20} \to pN_{22}p$	0.016	0.924
$N_{10} \to N_{13}$	1.0	1.0	$N_{20} \to pN_{23}p$	8.1×10^{-8}	3.7×10^{-8}
$N_{13} \to N_{14}$	3.6×10^{-8}	4.8×10^{-8}	$N_{20} \to pN_{24}p$	0.492	0.038
$N_{13} \to N_{15}$	0.999	0.999	$N_{21} \to bN_{23}$	0.747	0.601
$N_{14} \to e$	1.0	1.0	$N_{21} \to bN_{21}$	0.253	0.398
$N_{15} \to a$	1.0	1.0	$N_{23} \to N_8$	1.0	1.0

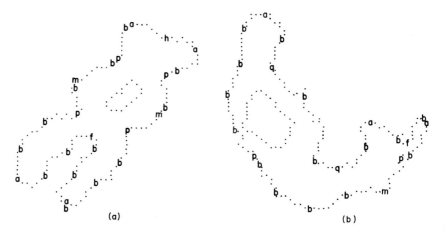

Fig. 6.9. (a) Feature-encoded abnormal chromosome. (b) Feature-encoded normal chromosome.

Typical sample chromosome patterns labeled by primitives are shown in Fig. 6.9. The grammar $\bar{G}_i$ is already in Wirth–Weber precedence form. By applying the stochastic syntax analysis procedure described in Section 5.5, the following classification results are obtained [6].

Sample string	Machine classification	Actual class
fbebqzqbbbbebbpbzbbbpbeb	Normal	Normal
fbabbbqbmbbqbagabqzbqbbbbabb	Normal	Normal
fbbabbpbmbpbahabpbmpbbbabb	Abnormal	Abnormal
fbabpmbbbbpbbbbabqbzbqbab	Normal	Normal
fabbbbpmbpbbahabpbbmbpbab	Normal	Normal
fbabpbzbbbpakagabpbbbzpbab	Abnormal	Abnormal
fbbaqmbqbbbabbbbpmbbbpbab	Normal	Normal
fbbabbpbzpbabbpbbmbbbpbbab	Normal	Normal
fbabbbbbqzqagabbpzbbpbabb	Normal	Normal
fbabqzqbbbbabgbaqbzbqbbab	Normal	Normal
fbbabbbqbmqabgbabbqbmbqbab	Normal	Normal
fbabbqbzbqbabgbabqbzbqbbabb	Normal	Normal
fbbaqzbbqbbbabgaqmqbbbbabb	Normal	Normal
fbbabbbbmbpbagabpmpbbbabb	Normal	Normal
fbbabbqmbbqagaqmbbqbbbabb	Abnormal∗	Normal
fbaqbzbqbbbaqmbbbqbbbab	Normal	Normal

∗Misclassification.

The behavior of stochastic syntax analyzers for stochastic cfpg with a fixed (or pure) strategy and a random strategy (Section 5.6) is demonstrated in Examples 6.8 and 6.9, respectively. Two different scfpg's are used to evaluate the performance of the stochastic syntax analyzers [9].

Let G_{sp} be a consistent stochastic cfpg. The average number of steps needed to recognize the language $L(\bar{G}_{sp})$ is defined as

$$M = \sum_{x \in L(\bar{G}_{sp})} p(x)n(x)$$

where $p(x)$ is the probability of a string x being generated by G_{sp} and $n(x)$ is the number of steps required by the syntax analyzer to recognize the string x. The empirical average number of steps to recognize the set of N strings $\{x_1, \ldots, x_N\}$ that are statistically, independently generated by G_{sp} is defined as

$$M_N = \left(\sum_{i=1}^{N} n(x_i) \right) \bigg/ N$$

where M_N approaches M as $N \to \infty$.

Example 6.8   Consider an scfpg describing block Arabic numerals: $G_1 = (V_N, V_T, J, P_s, S)$, where $V_N = \{S, A, B, C, D, E, F\}$, $V_T = \{a, b, c, d, e, f, g, h\}$, $J = \{1, 2, \ldots, 27\}$, and P_s is defined as follows:

Label	Core	U	P(U)
1	$S \rightarrow CDAB$	2, 3	0.9, 0.1
2	$C \rightarrow cF$	3, 4	0.77, 0.23
3	$F \rightarrow c$	5, 7	0.71, 0.29
4	$F \rightarrow g$	6, 8	0.5, 0.5
5	$D \rightarrow d$	9, 13, 20	0.6, 0.2, 0.2
6	$D \rightarrow d$	10	1.0
7	$D \rightarrow h$	11, 20	0.5, 0.5
8	$D \rightarrow h$	12	1.0
9	$A \rightarrow aE$	14, 17	0.67, 0.33
10	$A \rightarrow aE$	16	1.0
11	$A \rightarrow aE$	17	1.0
12	$A \rightarrow aE$	15	1.0
13	$A \rightarrow eE$	16	1.0
14	$E \rightarrow ab$	18, 22	0.5, 0.5
15	$E \rightarrow ab$	19	1.0
16	$E \rightarrow eb$	19	1.0
17	$E \rightarrow b$	22	1.0
18	$B \rightarrow gd$	$\varnothing$	1.0
19	$B \rightarrow cd$	$\varnothing$	1.0
20	$A \rightarrow e$	21	1.0
21	$B \rightarrow f$	$\varnothing$	1.0
22	$B \rightarrow c$	$\varnothing$	1.0
23	$C \rightarrow gF$	24	1.0
24	$F \rightarrow c$	$\varnothing$	1.0
25	$D \rightarrow d$	26	1.0
26	$A \rightarrow eE$	27	1.0
27	$E \rightarrow ab$	18, 22	0.5, 0.5

where $\{a, b, c, d, e, f, g, h\}$ is the set of primitives shown in Fig. 6.10a. The set of patterns generated by G_1 is the set of block Arabic numerals shown in Fig. 6.10b. Each pattern shown in Fig. 6.10b can be generated with almost equal probability.†

Let G_2, G_3, G_4, G_5, and G_6 be scfpg obtained from G_1 by rearranging the order of the alternatives defined in each go-to field of G_1. The performance of the nondeterministic syntax analyzers constructed according to $G_1, \ldots, G_6$ and the performance of the stochastic syntax analyzer with a random strategy

† The scfpg considered in this example was constructed by Mr. B. K. Bhargava of Purdue University.

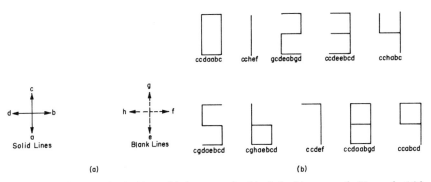

Fig. 6.10. (a) A set of primitives; (b) the respective block Arabic numerals (Example 6.8.)

are shown in Fig. 6.11. The nondeterministic syntax analyzer constructed according to G_1 is the stochastic syntax analyzer with a fixed strategy. In these simulations, 250 strings are statistically, independently generated by G_1.

Example 6.9 Consider a scfpg describing rectangles:

$$G_1 = (V_N, V_T, J, P_s, S)$$

where $V_N = \{S, A, B, C, D\}$, $V_T = \{a, b, c, d\}$, $J = \{1, 2, \ldots, 9\}$, and P_s is

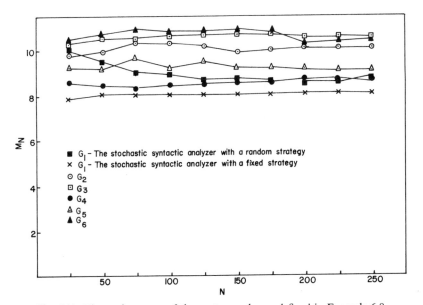

Fig. 6.11. The performance of the syntax analyzers defined in Example 6.8.

defined as follows:

Label	Core	U	$P(U)$
1	$S \rightarrow ABCD$	2, 4, 6	0.4, 0.4, 0.2
2	$A \rightarrow aA$	3	1.0
3	$C \rightarrow cC$	2, 4	0.7, 0.3
4	$B \rightarrow bB$	5	1.0
5	$D \rightarrow dD$	4, 6	0.5, 0.5
6	$A \rightarrow a$	7	1.0
7	$B \rightarrow b$	8	1.0
8	$C \rightarrow c$	9	1.0
9	$D \rightarrow d$	$\varnothing$	1.0

where $\{a, b, c, d\}$ is the set of primitives shown in Fig. 6.12a. It is simple to show that G_1 defines a set of rectangles some of which are shown in Fig. 6.12b.

Let G_2, G_3, G_4, G_5, and G_6 be scfpg obtained from G_1 by rearranging the order of the alternatives defined in each go-to field of G_1. The performance of each nondeterministic syntax analyzer constructed according to $G_1, \ldots, G_6$

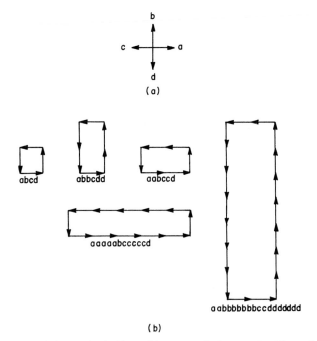

(a)

(b)

Fig. 6.12. (a) A set of primitives; (b) some typical rectangles (Example 6.9).

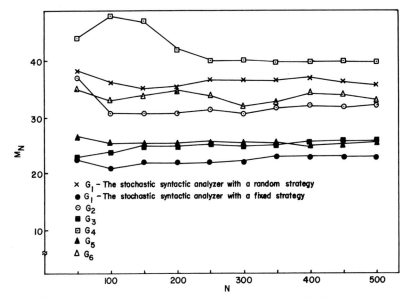

Fig. 6.13. The performance of the syntax analyzers defined in Example 6.9.

and the performance of the stochastic syntax analyzer with a random strategy are shown in Fig. 6.13. The stochastic analyzer with a fixed strategy is the nondeterministic analyzer constructed according to G_1.

The simulation results presented in Examples 6.8 and 6.9 clearly support the conjecture that the stochastic syntax analyzer with a fixed strategy would require the least average number of steps to recognize the language defined by the given scfpg, and the performance of the stochastic syntax analyzer with a random strategy is no worse than that of some nondeterministic syntax analyzer with regard to reducing the average number of steps needed to recognize the language generated by the given scfpg.

Example 6.10 The stochastic syntax analysis procedure with a fixed strategy (Section 5.6) is applied to the classification of chromosome patterns based on the three grammars described in Example 6.4. The maximum-likelihood recognizer shown in Fig. 6.4 is used. The following set of production probabilities is assigned:

$$p_{1m} = 0.9, \quad p_{2m} = 0.7, \quad p_{3m} = 0.8, \quad p_{4m} = 0.7$$

$$p_{1s} = 0.9, \quad p_{2s} = 0.7, \quad p_{3s} = 0.3, \quad p_{4s} = 0.6, \quad p_{5s} = 0.3, \quad p_{6s} = 0.8,$$

$$p_{7s} = 0.7$$

$$p_{1a} = 0.4, \quad p_{2a} = 0.4, \quad p_{3a} = 0.5, \quad p_{4a} = 0.8, \quad p_{5a} = 0.7$$

One hundred sample strings are generated statistically by each of the three chromosome grammars. Classification results are shown in the following [9].

	Median	Submedian	Acrocentric
Median	99	0	1
Submedian	3	90	7
Acrocentric	0	0	100

References

1. U. Grenander, *Syntax-Controlled Probabilities*. Tech. Rep. Div. of Appl. Math., Brown Univ., Providence, Rhode Island, 1967.
2. K. S. Fu and P. H. Swain, On syntactic pattern recognition. In *Software Engineering* (J. T. Tou, ed.), Vol. 2. Academic Press, New York, 1971.
3. K. S. Fu, Syntactic pattern recognition and stochastic languages. In *Frontiers of Pattern Recognition* (S. Watanabe, ed.). Academic Press, New York, 1972.
4. P. H. Swain and K. S. Fu, Stochastic programmed grammars for syntactic pattern recognition. *Pattern Recognition* **4**, 83–100 (1972).
5. H. C. Lee and K. S. Fu, *Stochastic Linguistics for Picture Recognition*. Rep. TR-EE 72-17. School of Elec. Eng., Purdue Univ., West Lafayette, Indiana, 1972.
6. H. C. Lee and K. S. Fu, A syntactic pattern recognition system with learning capability. *Int. Symp. Comput. Inform. Sci. 4th, Miami Beach, Florida, December 14–16, 1972.*
7. T. W. Anderson and L. A. Goodman, Statistical inference about Markov chains. *Ann. Math. Statist.* **28**, 89–109 (1957).
8. W. G. Cochran, The χ^2-test of goodness of fit. *Ann. Math. Statist.* **27**, 1–22 (1956).
9. T. Huang and K. S. Fu, Stochastic syntax analysis for programmed grammars and syntactic pattern recognition. *Comput. Graphics and Image Process.* **1**, 257–283 (1972).
10. V. D. Dimitrov, Multi-layered stochastic languages and stochastic grammars for syntactic pattern recognition. *Int. Symp. Theor. Problems and Syst. for Pattern and Situation Recognition, Varna, Bulgaria, October, 8–12, 1972.*

Chapter 7

Grammatical Inference for Syntactic Pattern Recognition

7.1 Introduction and Basic Definitions

The use of formal linguistics in modeling natural and programming languages and in describing physical patterns and data structures has recently received increasing attention [1–6]. Grammars or syntax rules are employed to describe the syntax of languages or the structural relations of patterns. In addition to the structural description, a grammar can also be used to characterize a syntactic source which generates all the sentences (finite or infinite) in a language, or the patterns belonging to a particular class. In order to model a language or to describe a class of patterns or data structures under study more realistically, it is hoped that the grammar used can be directly inferred from a set of sample sentences or a set of sample patterns. This problem of learning a grammar based on a set of sample sentences is called grammatical inference.† Potential engineering applications of grammatical inference include areas of pattern recognition, information retrieval, programming language design, translation and compiling, graphics languages, man–machine communication, and artificial intelligence [6, 14–21].

† For other related approaches in the learning of pattern structures, see the literature [7–13].

Formally speaking, the problem of grammatical inference is concerned mainly with the procedures that can be used to infer the syntactic rules of an unknown grammar G based on a finite set of sentences or strings S_t from $L(G)$, the language generated by G, and possibly also on a finite set of strings from the complement of $L(G)$. The inferred grammar is a set of rules for describing the given finite set of strings from $L(G)$ and predicting other strings which in some sense are of the same nature as the given set. A basic block diagram of a grammatical inference machine is shown in Fig. 7.1. The measure of goodness

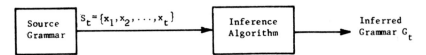

Fig. 7.1. A basic model of a grammatical inference machine.

of the inferred grammar for S_t can be arbitrarily defined such that in some meaningful sense it yields a satisfactory result. In recent years some measures of goodness have been defined in terms of the complexity of the inferred grammar, and they have been applied to grammatical inference problems.

We shall first establish the definitions and notations that will be used throughout this chapter, and present a brief review of some major results in grammatical inference and some examples to illustrate the applications to syntactic pattern recognition [22–42].

Definition 7.1   An information sequence of a language L, denoted by $I(L)$, is a sequence of strings from the set $\{+y \,|\, y \in L\} \cup \{-y \,|\, y \in V_T^* - L\}$. A positive information sequence of a language L, denoted by $I^+(L)$, is an information sequence of L containing only strings from L. Similarly, a negative information sequence of a language L, denoted by $I^-(L)$, is an information sequence of L containing only strings from $V_T^* - L$.

Definition 7.2   An information sequence of a language L is said to be complete if (i) $I^+(L)$ contains all strings from L and (ii) $I^-(L)$ contains all strings not in L. A positive information sequence $I^+(L)$ is said to be complete if each string in L appears in $I^+(L)$.

Definition 7.3   A sample of a language L, denoted by $S_t(L)$, is defined to be the set $\{+y_1, \ldots, +y_t\} \cup \{-y_1, \ldots, -y_t\}$ where $\{+y_1, \ldots, +y_t\}$ is defined to be the positive sample and $\{-y_1, \ldots, -y_t\}$ the negative sample. A positive sample S_t of a language $L(G)$ is structurally complete if each production defined in G is used in the generation of at least one string in S_t.

Definition 7.4   Let $z \in V_T^*$. Then the k tail of z with respect to $S \in 2^{V_T^*}$, denoted by $g(z, S, k)$, is defined as follows:

$$g(z, S, k) = \{x \in V_T^* \,|\, zx \in S \text{ and } |x| \le k\}$$

where $|x|$ denotes the length of the string x, and $k > 0$. $g(z, S, k)$ is not defined if there is no string $x \in V_T^*$ such that $zx \in S$ and $|x| \leq k$.

Definition 7.5 Let f be a function that maps the nonterminal set of grammar G_2 into the nonterminal set of grammar G_1. In addition, suppose that f maps the initial nonterminal of G_2 to the initial nonterminal of G_1 and f is applied to map productions of G_2 onto the productions of G_1. Then G_1 will be said to cover G_2, and furthermore, $L(G_2) \subseteq L(G_1)$ [43].

Definition 7.6 A pivot grammar is a grammar $G_p = (V_N, V_T, P, S)$ where V_N is a finite set of nonterminals; V_T is a finite set of terminals that can be partitioned into two disjoint sets, denoted by V_{Tp} and V_{T0} ; P is a finite set of productions of the following forms: (1) $A \rightarrow BaC$; (2) $A \rightarrow Bb$; (3) $A \rightarrow bB$; and (4) $A \rightarrow b$; where $A, B, C \in V_N$; $a \in V_{Tp}$, and $b \in V_{T0}$; S is the start symbol.

Definition 7.7 A grammar G_t is said to be compatible with a sample S_t if $L(G_t)$ contains all strings in the positive sample, and none of the strings in the negative sample.

Definition 7.8 A class of grammars C is said to be admissible if (i) C is denumerable, and (ii) for $x \in V_T^*$, it is decidable whether or not $x \in L(G)$ for any $G \in C$.

Example 7.1 The classes of context-sensitive, context-free, and regular grammars are all denumerable and recursive. Hence, they are all admissible.

Definition 7.9 A class of languages $L(G)$ is said to be identifiable in the limit if there is an inference algorithm D such that for any $G \in C$ and any complete information sequence $I(L(G))$ there exists a τ and such that

$$\text{(i)} \quad G_t = G_\tau \quad \text{for} \quad t > \tau \tag{7.1}$$

where $G_t = D(S_t, C)$ and $G = D(S_\tau, C)$, and

$$\text{(ii)} \quad L(G_\tau) = L(G) \tag{7.2}$$

Definition 7.10 The algorithm D is said to approach the grammar G if the following two conditions hold. (a) For any $x \in L(G)$ there is a τ such that $t > \tau$ implies $x \in L(G_t)$, where $G_t = D(S_t, C)$. (b) For any G' such that $L(G') - L(G) \neq \emptyset$ there is a τ such that $t > \tau$ implies $G_t \neq G'$.
D is said to strongly approach G if, in addition to (a) and (b), there is an H such that $L(H) = L(G)$ and for any infinite number of t, $G_t = H$.

Intuitively, an algorithm D can be used to identify G if it eventually guesses only one grammar and that grammar generates exactly $L(G)$. This does imply that D can effectively choose one grammar and stop considering new data. If D can only be guaranteed to eventually reject any grammar not producing $L(G)$, then $L(G)$ is approachable.

Definition 7.11   A context-free grammar (cfg) $G = (V_N, V_T, P, S)$ is said to be completely reduced if (i) P contains no production of the form $A \rightarrow \lambda$ for $A \in V_N - S$. (ii) P contains no production of the form $A \rightarrow B$ for $A, B \in V_n$. (iii) If $S \overset{*}{\Longrightarrow} \alpha$, $\alpha \in (V_N \cup V_T)^*$, there is an $x \in V_T^*$ such that $\alpha \overset{*}{\Longrightarrow} x$. (iv) Each production is used in the generation of at least one string in $L(G)$.

Definition 7.12   Let $G = (V_N, V_T, P, S)$ be a context-free grammar. The parenthesis grammar of G, denoted by $[G]$, is formed from G by replacing every production $A \rightarrow \alpha$ by $A \rightarrow [\alpha]$ where [and] are special symbols not in V_T.

Definition 7.13   Let $G = (V_N, V_T, P, S)$ be a completely reduced cfg. A structural information sequence of $L(G)$, denoted by $I_s(L)$, is a sequence of strings from the set $\{+y \mid y \in L([G])\} \cup \{-y \mid y \in (V_T \cup \{[,]\})^* - L([G])\}$. A positive structural information sequence, denoted by $I_s^+(L)$, is a sequence of strings from $L([G])$. Similarly, a sequence of strings not in $L([G])$ is a negative structural information sequence.

Definition 7.14   Let G be a completely reduced cfg. A positive structural information sequence $I_s^+(L(G))$ is complete if every string in $L([G])$ appears at least once in $I_s^+(L(G))$.

Definition 7.15   For a grammar $G = (V_N, V_T, P, S)$, let x be a nonempty string on $V = V_N \cup V_T$. The leftmost and rightmost terminal sets of x, denoted as $L_t(x)$ and $R_t(x)$, are defined respectively as:

$$L_t(x) = \{a \mid x \overset{*}{\underset{G}{\Longrightarrow}} a\alpha \quad \text{or} \quad (x \overset{*}{\underset{G}{\Longrightarrow}} A\beta \quad \text{and} \quad a \in L_t(A))\} \tag{7.3}$$

$$R_t(x) = \{a \mid x \overset{*}{\underset{G}{\Longrightarrow}} \alpha a \quad \text{or} \quad (x \overset{*}{\underset{G}{\Longrightarrow}} \beta A \quad \text{and} \quad a \in R_t(A))\} \tag{7.4}$$

where $a \in V_T$, $A \in V_N$, and $\alpha, \beta \in V^*$, Intuitively, $L_t(x)$ ($R_t(x)$) consists of the terminals which are the leftmost (rightmost) in some derivation of x.

Definition 7.16   Let $G = (V_N, V_T, P, S)$ be a completely reduced cfg. The left and right profile of order k of a string x in $(V_N \cup V_T \cup \{[,]\})^+$, denoted by $L_k(x)$ and $R_k(x)$, respectively, are defined as follows. Let $V_T' = V_T \cup \{[,]\}$,

$$L_k(x) = \left\{ u \mid x \overset{*}{\underset{[G]}{\Longrightarrow}} y_1 \cdots y_m, y_i \in V_T', u = \begin{cases} y_1 \cdots y_k, & \text{if } m \geq k \\ y_1 \cdots y_m \$^{k-m} & \text{if } m \leq k \end{cases} \right\}$$

$$\tag{7.5}$$

and

$$R_k(x) = \left\{ u \,\middle|\, x \xRightarrow[{[G]}]{*} y_1 \cdots y_m , y_i \in V_T', u = \begin{cases} y_{m-k+1} \cdots y_m & \text{if } m \geq k \\ \$^{k-m} y_1 \cdots y_m & \text{if } m \leq k \end{cases} \right\}$$

(7.6)

where $\$ \notin V_T'$.

Definition 7.17   Let $G = (V_N, V_T, P, S)$ be a completely reduced cfg. The k profile of a string $x \in (V_N \cup V_T \cup \{[\,,\,]\})^+$, denoted by $P_k(x)$, is

$$P_k(x) = (L_k(x); R_k(x))$$

(7.7)

Definition 7.18   A completely reduced cfg $G = (V_N, V_T, P, S)$ is k distinct if for any two distinct nonterminals A and B in V_N, $P_k(A) \neq P_k(B)$.

Definition 7.19   Let $G = (V_N, V_T, P, S)$ be a completely reduced cfg. G is k homogeneous if for any two distinct productions of the form $A \to x$ and $A \to y$, $P_k(x) = P_k(y)$.

Definition 7.20   A positive stochastic information sequence of a stochastic language $L_s = L(G_s)$, denoted by $I(L_s)$, is a sequence of strings each of which is a random variable statistically generated by the stochastic grammar G_s. A positive stochastic sample of a stochastic language $L_s = L(G_s)$, denoted by $S_t(L_s)$, is defined as the set $\{s_1, \ldots, s_t\}$ where $s_1, \ldots, s_t$ are strings (random variables) generated by the stochastic grammar G_s.

The various grammatical inference algorithms developed so far can be classified into two distinct categories: (1) grammatical inference by enumeration; and (2) grammatical inference by induction. They are briefly reviewed in the next three sections.

7.2   Grammatical Inference Algorithms by Enumeration

7.2.1   Basic Formulation

Gold [22] has presented a study of language identification in the limit by enumeration. His model of language identification consists of three basic components:

(1)   A denumerable class C of grammars. Suppose that some grammar $G \in C$ is chosen. A grammatical inference machine is to determine, based on an information sequence of the language L generated by G, which grammar it is in C.

(2)   A method of information presentation. Assume that time is quantized and start at a finite time, that is, $t = 1, 2, 3, \ldots$. At each time t, the grammatical

inference machine is presented with a string that is chosen from an information sequence of L. Text presentation and informant presentation are the two basic methods of information presentation considered. An informant presentation of L is a complete information sequence of L while a text presentation of L is a complete positive information sequence of L.

(3) A grammatical inference algorithm. A grammatical inference algorithm D can form a guess on the basis of S_t such that $D(S_t, C) = G_t$ for $G_t \in C$. All algorithms are constructed based on the denumerability of various classes of grammars.

The main results of Gold's study in language identification in the limit are concerned with the distinct difference in identifiability of a language generated by a grammar in an admissible class of grammars effected by two methods of information presentation, that is, text presentation and informant presentation. Gold's results on language identification in the limit by enumeration and some of their extension by Feldman [24] are summarized into the following theorems.

Theorem 7.1 Let C be an admissible class of grammars. There is an algorithm such that for any $G \in C$, $L(G)$ is identifiable in the limit through an informant presentation [22].

Theorem 7.2 Let C be an admissible class of grammars which generates all *finite* languages and any one infinite language L. Then L is not identifiable in the limit through a text presentation [22].

Theorem 7.3 Let C be a class of regular languages. C is not identifiable in the limit through a text presentation [22].

Theorem 7.4 For any admissible class C of grammars, there is an algorithm D such that for any $G \in C$, G is strongly approachable through a text presentation [24].

The works by Gold and Feldman provide a basic formulation for theoretical study in grammatical inference by enumeration. Various methods have been suggested to improve the enumerative efficiency by eliminating large classes of grammars from the search [35]. Often, a particular application would provide additional information about the desired form (structural information) of grammars. Under such a condition, only grammars of the desired form need to be considered, and they can usually be enumerated in a fashion that eliminates the generation of an unnecessarily large number of equivalent grammars. Also, it is known that the only grammars that need to be examined are those that are compatible with the given sample, and many incompatible grammars can be eliminated without ever being created. This elimination can usually be carried out by deleting classes of incompatible grammars on the basis of a test on one of them. The concept of grammatical covering enables us to discard all grammars which are covered by G if G does not generate all the known strings in the language because the covered grammars will have the same fault. On the other hand, if G generates some strings that are known to be outside the language, then all grammars which cover G will be similarly inadequate. Consequently, the discovery that a particular grammar in an enumeration is not compatible with the given sample makes it possible to eliminate from consideration that grammar and possibly a whole class of other grammars.

7.2.2 Grammatical Inference Via a Lattice Structure

Pao [28] has developed a finite search algorithm to infer a finite-state grammar for a positive sample S_t. The strategy is first to construct a finite-state grammar which will generate exactly those strings in S_t. By examining all possible combinations of nonterminals with the help of an informant, the algorithm constructs a finite-state grammar which will generate all strings in S_t and possibly some other strings. If the positive sample is structurally complete, the source grammar, a finite-state grammar of course, can be identified in the limit.

Suppose $\{Z_1, \ldots, Z_n\}$ is the set of nonterminals which appear in the finite-state grammar which generates exactly those strings in S_t. All possible partitions of the nonterminal set are considered. The corresponding grammar of a partition is formed by merging all nonterminals in each equivalence class into a single nonterminal. It has also been shown [44] that the set of all partitions and corresponding grammars can be ordered in a lattice. The grammar associated with any node on the lattice will cover the grammars associated with its connected lower nodes on the lattice, and the grammar covering concept as a pruning technique in the enumeration is applicable. The algorithm for identifying the correct grammar can be briefly described as follows.

(1) Construct grammars G_1 and G_2, corresponding to the two lowest partitions in the lattice.

(2) Determine if G_1 is equivalent to G_2. If G_1 is equivalent to G_2, the partition corresponding to G_1 is eliminated, and go to (1). If the partition corresponding to G_2 is the only partition left in the lattice, G_2 is the solution.

(3) If G_1 and G_2 are not equivalent, construct $G_{12} = G_1 - G_2$ which generates strings in $L(G_1)$ but not in $L(G_2)$. Take a string x in $L(G_{12})$ and ask the informant whether x is a valid string (refer to Fig. 7.2). By a valid string we mean that the string can be generated by the source grammar. If x is a valid string, the partition corresponding to G_2 is eliminated from the

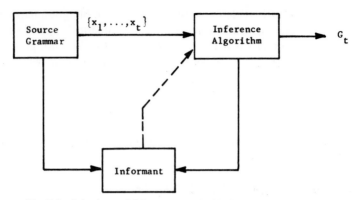

Fig. 7.2. A basic model for grammatical inference with informant.

lattice. If there is no partition left in the lattice, G_1 is the solution. Otherwise, go to (1).

(4)   If x is not a valid string, the partition corresponding to G_1 is eliminated from the lattice. If there is no partition left in the lattice, G_2 is the solution. Otherwise go to (1).

Instead of considering the class of finite-state grammars, Pao limits the hypothesis space for her grammatical inference model to a finite set of finite-state grammars by requiring that S_t be structurally complete. It appears that if the number of strings in S_t were large, the number of partitions to be considered would become astronomical. If the positive sample is not structurally complete, it will not be possible to identify the source grammar. Hence, the initial choice of sample is critical. The main result obtained from Pao's algorithm can be stated in the following theorem.

Theorem 7.5   Let C be a class of finite-state grammars. For any G in C, G is identifiable in finite time through a structurally complete positive sample of $L(G)$ and an informant.

7.2.3   Grammatical Inference Using Structural Information Sequence

Crespi-Reghizzi has proposed an inference algorithm for operator precedence grammars from a structural information sequence $I_s(L)$ of a language L [30]. His approach is to infer an operator precedence grammar that is compatible with the given sample S_t of an operator precedence language L. An example is used here to illustrate the inference procedure.

Example 7.2   Assume that L is the language of arithmetic expressions such as $a + a, a + (a + a), \ldots$ on the terminals $V_T = \{a, +, (,)\}$. Let the structured sample $S_1 = \{s_1\}$ be

$$s_1 = [[a] + [[a] + [a]]] \tag{7.8}$$

Square brackets delimit constituents in the string of the sample, in an order that corresponds to evaluation of the string from right to left. In order to construct a grammar compatible with S_1, the inference algorithm scans the string s_1 from left to right and locates the first substring enclosed by the innermost brackets. In this example, the substring $[a]$ is found. Any grammar, in order to generate s_1 with the given structure, must have a production $N \to a$ where N represents a nonterminal. The algorithm therefore assigns to N, in the production $N \to x$, $x \in V^+$, the name consisting of the leftmost and rightmost terminal sets of x, that is, the order pair $(L_t(x), R_t(x))$. The order pair $(L_t(x), R_t(t))$ is usually written as $\langle L_t(x)NR_t(x)\rangle$. In the example, $x = a$, $L_t(a) = \{a\}$, $R_t(a) = \{a\}$, and we write the production $N \to a$ as $\langle aNa\rangle \to a$. Using this production, the string s_1 reduces to

$$s_1 = [\langle aNa\rangle + [\langle aNa\rangle + \langle aNa\rangle]] \tag{7.9}$$

The innermost brackets imply the production $N \to \langle aNa\rangle + \langle aNa\rangle$, which is rewritten, in accordance with the naming rule, as $\langle a + Na +\rangle \to \langle aNa\rangle + \langle aNa\rangle$ since $L_t(\langle aNa\rangle + \langle aNa\rangle) = L_t(\langle aNa\rangle) \cup \{+\} = \{a, +\}$ and $R_t(\langle aNa\rangle + \langle aNa\rangle) = \{+\} \cup R_t(\langle aNa\rangle) = \{a, +\}$. The string s_1 now reduces to

$$s_1 = [\langle aNa\rangle + \langle a + Na +\rangle] \tag{7.10}$$

which requires in turn the production $N \to \langle aNa\rangle + \langle a + Na +\rangle$ to be written as $\langle a + Na +\rangle \to \langle aNa\rangle + \langle a + Na +\rangle$. The string s_1 eventually becomes $[\langle a + Na +\rangle]$ and this nonterminal is renamed S with the production $S \to \langle a + Na +\rangle$. Thus, a grammar G_1 has been constructed which consists of productions

$$
\begin{array}{lll}
G_1: & A \to a & A \equiv \langle aNa\rangle \\
& B \to A + A & B \equiv \langle a + Na +\rangle \\
& B \to A + B & \\
& S \to B &
\end{array}
$$

The language $L(G_1) = \{a + a + a,\ a + a + a + a,\ \ldots\}$ is compatible with the sample S_1 since $L([G_1]) \supset S_1$. It is noted that G_1 also generates infinitely many strings which are a subjectively reasonable generalization of the given sample. When a second string

$$s_2 = [[([a] + [a])] + [a]] \tag{7.11}$$

is added to the previous sample S_1, we have a new sample $S_2 = \{S_1 \cup \{s_2\}\} = \{s_1, s_2\}$. The algorithm examines s_2 and produces the following productions.

$$
\begin{array}{lll}
F_2: & A \to a & A \equiv \langle aNa\rangle \\
& B \to A + A & B \equiv \langle a + Na +\rangle \\
& C \to (B) & C \equiv \langle (N)\rangle \\
& D \to C + A & D \equiv \langle +(Na +\rangle \\
& S \to D &
\end{array}
$$

It is obvious that F_2 is compatible with $\{s_2\}$; that is, $L([F_2]) \supset \{s_2\}$. In order to construct a grammar G_2 compatible with S_2, the algorithm performs the union of G_1 and F_2.

$$G_2 = G_1 \cup F_2 :$$

$$A \to a$$

$$B \to A + A \qquad B \to A + B$$

$$C \to (B)$$

$$D \to C + B$$

$$S \to B \qquad\qquad S \to D$$

where $A \equiv \langle aNa \rangle$, $B \equiv \langle a + Na + \rangle$, $C \equiv \langle (N) \rangle$, and $D \equiv \langle +(Na +\rangle$. It can be verified that

$$L(G_2) = L(G_1 \cup F_2) \supset L(G_1) \cup L(F_2) \tag{7.12}$$

and that $L(G_2) = \{a + a, a + a + a, \ldots\} \cup \{(a + a) + a, (a + a + a) + a, \ldots\}$.

On the basis of the two strings $\{s_1, s_2\} = S_2$, the algorithm does not infer the complete class of arithmetic expressions. For example, G_2 fails to generate $a + (a + a)$, $a + (a + a) + a$, etc. The sample S_2 is, therefore, not a representative of the language, and more strings need to be added in order to identify the source language. Suppose the following strings are added to the sample S_2. (Brackets are omitted for the sake or readability.)

$$s_3 = ((a + a + a)) + a + (a + a), \qquad s_7 = (a + a) + (a + a) + a$$

$$s_4 = (a + (a + a) + a), \qquad\qquad s_8 = a + a + (a + a)$$

$$s_5 = ((a + a) + ((a + a) + a)), \qquad s_9 = (a + (a + a))$$

$$s_6 = a((a + a) + a), \qquad\qquad s_{10} = a + (a + a) + (a + a) + (a + a)$$

Let $S_{10} = S_2 \cup \{s_3, \ldots, s_{10}\}$. The algorithm will construct the grammar

$$G_{10}:$$

$A \to a,$			where $A \equiv \langle aNa \rangle$
$B \to A + A,$	$B \to A + B,$	$B \to A + D,$	where $B \equiv \langle a + Na + \rangle$
$C \to (B),$	$C \to (C),$	$C \to (D),$	
$C \to (E),$	$C \to (F),$		where $C \equiv \langle (N) \rangle$
$D \to C + B,$	$D \to C + D,$	$D \to C + A,$	where $D \equiv \langle +(Na +\rangle$
$E \to A + C,$	$E \to A + E,$	$E \to A + F,$	where $E \equiv \langle a + N + \rangle$
$F \to C + E,$	$F \to C + C,$	$F \to C + F,$	where $F \equiv \langle +(N +) \rangle$
$S \to B,$	$S \to C,$	$S \to D,$	
$S \to E,$	$S \to F$		

The language $L(G_{10})$ consists of the arithmetic expressions over the terminals $V_T = \{a, +, (,)\}$, with the exclusion of a, (a), $((a))$, $\ldots$, and is an operator precedence language. If the strings a, (a), $((a))$, ... are considered valid, the following two strings should be added to S_{10}.

$$s_{11} = a, \qquad s_{12} = (a)$$

Then, the inferred grammar G_{12} would generate all the arithmetic expressions.

The grammar inferred by the algorithm is an operator precedence grammar in a normal form, which is characterized by the following properties: (i) There are no renaming rules in the grammar, except for the rules $S \rightarrow A$, $S \rightarrow B$, ... having S as the left part of productions. (ii) There are no repeated right parts in the productions. (iii) All the right parts of the same nonterminal have identical (L_t, R_t) pairs. (iv) Two distinct nonterminals may not have an identical (L_t, R_t) pair. The class of grammars characterized by (iv) is called free operator precedence grammars. It can be shown that the identification in the limit is possible through only a text presentation of structural information and that the grammar G_t inferred from a sample S_t generates the smallest language, among the compatible grammars satisfying (i), (ii), (iii), and (iv). In other words, the inferred grammar is, in some sense, the best fit for the sample in the class of free operator precedence grammars. In addition, it can also be proved that the inferred grammar is the simplest grammar among the compatible grammars satisfying (i), (ii), (iii), and (iv) in terms of the number of nonterminals and productions.

The algorithm has also been extended to include more general classes of languages, that is, to the inference of k-distinct and k-homogeneous context-free grammars [31]. The inference procedure is summarized in the following.

(1)   For each string s_i in a positive structural information sequence, construct a cfg G_i that exactly generates the string s_i by the algorithm just described for the inference of operator precedence grammars. That is, substitute a distinct nonterminal A_i for each substring x in s_i such that x contains no left or right parenthesis, and create the production $A_i \rightarrow x$. This process is repeated again and again until s_i contains a single nonterminal. This nonterminal is defined as the start symbol S.

(2)   For each G_i, combine all nonterminals having the same k profile. Let the resulting grammar be denoted by G_i'.

(3)   The final grammar is $G' = \bigcup_i G_i'$.

The basic result obtained from this procedure can be stated in the following theorem.

Theorem 7.6   Let C be a class of k-distinct and k-homogeneous cfg. For any G in C, $L(G)$ is identifiable in the limit through a text presentation of structural information.

7.3   Grammatical Inference Algorithms by Induction

7.3.1   Grammar Acquisition through Inductive Inference

Solomonoff [32] has described a method by which to discover the nesting or recursive structure of a language L from a positive sample S_t. The strategy that he describes consists of the following steps.

(1)   Delete substrings of a valid string and ask the informant if the remaining string is acceptable (Fig. 7.2).

(2)   If the remaining string is acceptable, then we reinsert the deleted substring with several repetitions and ask if the resulting string is acceptable. If the resulting string is acceptable, a recursive construction is formed.

Example 7.3   Let aba be a string in S_t. The informant will be queried as to the validity of the following strings:

$$aa, \quad ab, \quad ba, \quad a, \quad b$$

Suppose the a's are deleted from a string aba in S_t and the remaining string b is a valid string. Then the informant is queried as to the validity of strings a^2ba^2, a^3ba^3, ..., a^kba^k. If they are all acceptable for a sufficiently large number k, a recursive construction $\{A \to aAa$ and $A \to b\}$ is formed.

It is clear that the choice of the initial sample S_t is extremely critical in discovering the recursive construction of a context-free language. If S_t is too small a set, not all the recursive construction of the language will be discovered, while if S_t is too large, the number of substrings to be considered becomes astronomically large. This method is not incremental with respect to S_t in the sense that the structure of the inferred grammar may be completely changed if more strings are added to S_t.

7.3.2   Inference of Finite-State Grammars

Feldman [23] has described a method for inferring a recursive, unambiguous finite-state grammar for a positive sample S_t. The strategy of his method is to construct first a nonrecursive grammar that generates exactly the given positive sample, and then to merge nonterminals in order to obtain a simpler, recursive finite-state grammar which will generate all strings in S_t and infinitely many other strings. The first part of Feldman's method can be simply described as follows. Suppose that $x = a_1a_2, \ldots, a_n$ is a string in S_t; then the set of productions $\{S \to a_1Z_1, Z_1 \to a_2Z_2, \ldots, Z_{n-2} \to a_{n-1}Z_{n-1}, Z_{n-1} \to a_n\}$ is constructed to generate exactly the string x. This simple procedure is carried out for each string in S_t. The grammar that generates exactly those strings in S_t is formed by taking the union of all productions formed in the foregoing manner for all strings in S_t. Then nonterminals are merged empirically to obtain a recursive finite-state grammar which generates all strings in S_t plus as few new (short) strings as possible.

Biermann and Feldman [33] have proposed two algorithms to synthesize a finite-state acceptor which accepts S_t and possibly some additional strings which resemble those of S_t based on a positive sample S_t.† It is clear that the hypothesis space for this grammatical inference model is the class of finite-state grammars. One of the two algorithms which construct the states and state transitions directly from S_t can be briefly described as follows. Let $S_t \in 2^{V_T^*}$ be the given positive sample of a finite-state language. The non-deterministic finite-state automaton $A(S_t, k) = (Q, \Sigma, f, Q_0, F)$ which accepts S_t and some additional strings can be constructed as follows.‡

$$Q = \{q \in 2^{V_T^*} | g(z, S_t, k) = q \text{ for some } z \in V_T^*\} \qquad (7.13)$$

in which $g(z, S_t, k)$ is defined in Definition 7.4.

$\Sigma = V_T$

$f(q, a) = \{q' \in Q | \text{ there is a } z \in V_T^* \text{ such that } g(z, S_t, k) = q \text{ and }$

$$g(za, S_t, k) = q'\} \qquad (7.14)$$

$Q_0 = \{g(\lambda, S_t, k)\}$, the initial state set, and

$F = \{q \in Q | \lambda \in q\}$, the set of final states.

The language inferred by this algorithm, though, is $T(A(S_t, k))$, the set of strings accepted by $A(S_t, k)$. Some basic results on the properties of $A(S_t, k)$ can be summarized as follows:

(1) $S \subseteq T(A(S_t, k))$ for $k > 0$.
(2) Let m be the length of the longest string in S_t. Then

$T(A(S_t, k)) = S_t$ for $k \geq m$ and $T(A(S_t, k)) = \Sigma^*$ for $k = 0$

(3) $T(A(S_t, k)) \supseteq T(A(S_t, k + 1))$ for $k \geq 0$.

It is evident that this algorithm has a simple operation, and is therefore easy to implement on a digital computer. Examples of applying this algorithm to the learning of grammars describing simple pictorial patterns can be found

---

† It should be noted that a reformulation of the problem in terms of the Nerode-like relations was presented in a recent paper by Biermann and Feldman [36] in order to simplify the theoretical proofs and generalize the results.

‡ Although the inference of a finite-state grammar is presented here by way of the synthesis of a finite-state automaton, a modification could be made on the inference algorithm to give a finite-state grammar directly.

in the work of Fu [38] and Lee and Fu [41]. The parameter k allows the user to obtain as exact a fit to the given sample S_t as he desires, by increasing the complexity of the structure of the resulting finite-state automaton (by adding more states). However, it may not be very easy to select k, if it is not known to the user, from S_t such that $T(A(S_t, k))$ provides a good fit of S_t. The algorithm is incremental; that is, if $A(S_t, k)$ is constructed to accept S_t and then S_t is changed slightly, only the states and state transitions in $A(S_t, k)$ that correspond to the changes in S_t need to be adjusted to yield a new finite-state automaton.

7.3.3   Inference of Pivot Grammars

Gips (see Feldman *et al.* [25]) has described a method by which to infer a pivot grammar that is an operator grammar of a very restricted form. The class of pivot grammars properly includes the class of linear grammars and is properly included within the class of context-free grammars.

The main strategy used by Gips to infer a pivot grammar for a positive sample S_t is to find the self-embedding in S_t. A distinct nonterminal is set aside as the recursive nonterminal. The inference algorithm can be carried out in two steps.

Step 1.   Each string in S_t is examined to see if it has a proper substring which appears in S_t. If it does not, it becomes a string in the working set. If it has a substring which appears in S_t, then the longest substring is replaced by the distinct nonterminal and the resulting string is placed in the working set. For example, consider $S_t = \{a - a, a - (a - a), (a - (a - a)) - a\}$. Since the string $a - a$ has no proper substring in S_t, the string $a - a$ is placed unchanged in the working set. The second string $a - (a - a)$ has a substring $a - a$ in S_t; hence the substring $a - a$ is replaced by the distinct nonterminal X and placed in the working set. Finally, since the third string $(a - (a - a)) - a$ has two substrings in S_t, namely, $a - a$ and $a - (a - a)$, only the longest substring, that is, $a - (a - a)$ is replaced by X and the resulting string is placed in the working set. After the first step of this algorithm is carried out, the working set contains $a - a$, $a - (X)$, and $(X) - a$.

Step 2.   A simple pivot grammar is constructed for the working set. For example, a pivot grammar for the working set above is

$$G: \quad X \to X_1 - X_2, \quad X_1 \to a,$$
$$X_1 \to (X_3, X_2 \to a, X_2 \to (X_4, X_3 \to X), X_4 \to X)$$

It can be seen from the description of this algorithm that the initial choice of S_t is quite critical.

7.3.4   A Heuristic Approach

Evans [20] has described a set of heuristic rules for the automatic inference of grammars which can be used for pattern description and recognition. The basic formulation for a pattern grammar consists of three basic items:

(1)   A set of primitives or pattern components of which the patterns under consideration are composed;

(2)   A set of predicates which describe the structure relationships among primitives defined in their arguments;

(3)   A set of productions, each of which consists of three essential parts:

   (i)   the name of the pattern being defined;

   (ii)   a list of variables (nonterminals) for the constituent parts of a pattern;

   (iii)   a condition that must be satisfied by the constituent parts in order that a pattern be defined.

Based on this formalism, a grammatical inference problem is defined as follows. Let $\{s_1, \ldots, s_n\}$ be the set of given patterns with a common syntactic structure. The problem is to produce, in terms of the available primitives and predicates, a pattern grammar that fits the given set of patterns. The set of heuristic rules for the inference of pattern grammars consists of the following.

(1)   For each sample s_i find all the structure that is imposed by the predicates of the primitives. Each time one or more predicates turns out true, a new "object" (subpattern) is created in the form of a production. Repeat the same procedure for the subpatterns and the primitives until nothing new can be built up.

(2)   Once all the pattern-building processes have been completed, there is a check to determine which of the subpatterns found have all the primitives of the samples as their constituent parts. All subpatterns that are not either in this set or among the constituent parts of a member of this set are discarded.

(3)   Choose a G_1 for describing s_1, a G_2 for s_2, and so on, and form

$$G = \bigcup_{i=1}^{n} G_i$$

for all possible ways. Apply the reduction rules in step (4) to obtain a modified pattern grammar.

(4)   The following reduction rules are used to modify the pattern grammar constructed in step (3) in order to obtain a "best" fit pattern grammar.

Rule 1.   If there are multiple occurrences of any production in G, eliminate all but one.

Rule 2.   Look for a pair of variables such that uniform substitution of one for the other throughout G would result in multiple occurrences of some productions. Make the substitution and apply Rule 1.

Rule 3.   Look for a pair (A, a) such that addition of the production $A \to a$ and some selective substitution of A for a in G would result in a reduction in the number of productions after elimination of multiple occurrences of productions.

Rule 4.   Set a threshold T_1 for the predicate-weakening procedure. If there are n productions, all of which would collapse into one production either directly or by an identification like that in Rule 2 or Rule 3 if we were to disregard any relational information contained in them, and if $n \geq T_1$, do the rewrite and then apply Rule 1.

Rule 5.   Set a threshold T_2 for the object(subpattern)-weakening procedure. If there are n productions that would be identical except for differences in primitive types (terminals) and $n \geq T_2$, rewrite the n rules as one rule, using the special name ANY, which means "any terminal type," and then apply Rule 1.

When no reduction is possible, the process stops.

Rules 4 and 5 provide some necessary transformations that work on primitives and predicates such that a set of productions can be collapsed into one in order to generate additional pattern descriptions that resemble those of samples.

After this process has been executed on each of the candidate grammars, resulting in a new set of candidate grammars, we must choose the "best" of them. One criterion suggested is based on the strength of a grammar. The figure of merit is defined as $F_G/(n_G)^2$, where F_G is the strength of the grammar G and n_G is the number of productions in G. The strength of a grammar G is simply the sum of the strength of its productions, and the strength of a production is defined as the sum of the number of its components that are primitives and the number of relation terms.

Example 7.4   The input pattern is the scene shown in Fig. 7.3a. The pattern primitives that occur are circle, dot, square, and line segment. Suppose that the available predicates are

$$\begin{array}{ll}
\text{LEFT}(X, Y) & \text{true if } X \text{ is to the left of } Y \\
\text{ABOVE}(X, Y) & \text{true if } X \text{ is above } Y \\
\text{INSIDE}(X, Y) & \text{true if } X \text{ is inside } Y
\end{array}$$

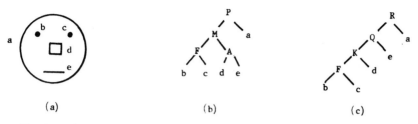

Fig. 7.3 (a) The scene and (b, c) the two structural descriptions in Example 7.4.

(1) First stage

$A \rightarrow$ ABOVE(d, e)
$B \rightarrow$ INSIDE(b, a)
$C \rightarrow$ INSIDE(c, a)
$D \rightarrow$ INSIDE(d, a)
$E \rightarrow$ INSIDE(e, a)
$F \rightarrow$ LEFT(b, c)
$G \rightarrow$ LEFT(b, c) $\wedge$ INSIDE(b, a) $\wedge$ INSIDE(c, a)
$H \rightarrow$ ABOVE(d, e) $\wedge$ INSIDE(d, a) $\wedge$ INSIDE(e, a)

(2) Second stage

$I \rightarrow$ INSIDE(A, a)
$J \rightarrow$ INSIDE(F, a)
$K \rightarrow$ ABOVE(F, d)
$L \rightarrow$ ABOVE(F, e)
$M \rightarrow$ ABOVE(F, A)

(3) Third stage

$N \rightarrow$ INSIDE(K, a)
$O \rightarrow$ INSIDE(L, a)
$P \rightarrow$ INSIDE(M, a)
$Q \rightarrow$ ABOVE(K, e)
$R \rightarrow$ INSIDE(Q, a)

P and R correspond to the two structural descriptions for the scene. They are shown represented as trees in Fig. 7.3b,c, respectively. The two grammars generating P and R are, respectively,

G_1: $R \rightarrow$ INSIDE(Q, circle)
        $Q \rightarrow$ ABOVE(K, lineseg.)
        $K \rightarrow$ ABOVE(F, square)
        $F \rightarrow$ LEFT(dot, dot)

G_2: $P \rightarrow$ INSIDE(M, a)
        $M \rightarrow$ ABOVE(F, A)
        $F \rightarrow$ LEFT(dot, dot)
        $A \rightarrow$ ABOVE(square, lineseg.)

Example 7.5   Given the sample

$$S_7 = \{caaab, bbaab, caab, bbab, cab, bbb, cb\}$$

The terminals are $a, b, c,$ and the only predicate is $CAT(X, Y)$, written as XY. For $s_1 = caaab,$

$$
\begin{aligned}
G_1: \quad & S \rightarrow cA \quad\quad B \rightarrow aC \\
& A \rightarrow aB \quad\quad C \rightarrow ab
\end{aligned}
$$

For $s_2 = bbaab,$

$$G_2: \quad S \rightarrow bbD, \quad\quad D \rightarrow aE, \quad\quad E \rightarrow ab$$

For $s_3 = caab,$

$$G_3: \quad S \rightarrow cF, \quad\quad F \rightarrow aG, \quad\quad G \rightarrow ab$$

For $s_4 = bbab,$

$$G_4: \quad S \rightarrow bbH, \quad\quad H \rightarrow ab$$

For $s_5 \rightarrow cab,$

$$G_5: \quad S \rightarrow cI, \quad\quad I \rightarrow ab$$

For $s_6 \rightarrow bbb,$

$$G_6: \quad S \rightarrow bbb$$

For $s_7 \rightarrow cb,$

$$G_7: \quad S \rightarrow cb$$

Thus

$$G = \bigcup_{i=1}^{7} G_i$$

The first set of reduction steps identifies $C, E, G, H,$ and I. The resulting grammar is

$$
\begin{aligned}
& S \rightarrow cA \quad\quad S \rightarrow bbD \quad\quad S \rightarrow bbC \\
& A \rightarrow aB \quad\quad D \rightarrow aC \quad\quad S \rightarrow cC \\
& B \rightarrow aC \quad\quad S \rightarrow cF \quad\quad S \rightarrow bbb \\
& C \rightarrow ab \quad\quad F \rightarrow aC \quad\quad S \rightarrow cb
\end{aligned}
$$

The next set of reduction steps identifies $A, C,$ and $F,$ resulting in:

$$
\begin{aligned}
& S \rightarrow cA \quad\quad S \rightarrow bbD \quad\quad S \rightarrow bbA \\
& A \rightarrow aB \quad\quad D \rightarrow aA \quad\quad S \rightarrow bbb \\
& B \rightarrow aA \quad\quad A \rightarrow aA \quad\quad S \rightarrow cb \\
& A \rightarrow ab
\end{aligned}
$$

Identifying A, B, and D results in

(1) $S \to cA$ (4) $S \to bbA$

(2) $A \to aA$ (5) $S \to bbb$

(3) $A \to ab$ (6) $S \to cb$

The next reduction introduces $A \to b$ and selectively substitutes A for b in productions (3), (5), and (6). We then have the final grammar:

$$S \to cA \qquad A \to aA$$
$$S \to bbA \qquad A \to b$$

7.4 Bayesian Inference of Stochastic Grammars

Horning [26, 27] has described an inference algorithm of stochastic grammars. Since the basic strategy of his inference algorithm is to choose the stochastic grammar G_i in a denumerable class C of stochastic grammars such that G_i maximizes the a posteriori conditional probability $P(G_i | S_t, C)$ of G_i when a given positive stochastic sample S_t is observed, it is called an enumerative Bayesian algorithm of stochastic grammars. Based on Gold's model for language identification, the inference model of Horning for stochastic grammars consists of three basic parts.

(1) A denumerable class C of stochastic grammars. A stochastic grammar $G_i \in C$ is chosen according to a given a priori probability $P(G_i | C)$ such that a grammatical inference machine is to determine, based on a positive stochastic information sequence of the stochastic language generated by G_i, which stochastic grammar it is in C.

(2) A stochastic text presentation. Assume that time is quantized and start at a finite time, that is, $t = 1, 2, 3, \ldots$. At each time t, the grammatical inference machine is presented with a string s_t which is a random variable generated by the source stochastic grammar $G_i \in C$ and $S_t = (s_1, \ldots, s_t)$ is a positive stochastic sample presented to the grammatical inference machine up to time t.

(3) A grammatical inference machine used in this model is called an enumerative Bayesian algorithm and can be described as follows. Let C be a denumerable class of stochastic grammars $(G_1, G_2, \ldots)$ with $T(\delta)$ a computable function such that $i > T(\delta)$ implies $P(G_i | C) < \delta$, and S_t be the given positive stochastic sample. The enumerative Bayesian algorithm consists of four basic steps.

(1) Let i_t be the least integer such that G_{i_t} is compatable with respect to S_t.

(2) Let $\delta_t = P(G_{i_t} | C)P(S_t | G_{i_t}, C)$.

(3) Let $i_T = T(\delta_t)$.
(4) For each G_i, $i_t \le i \le i_T$, compute $P(G_i|C)P(S_t|G_i, C)$. Let k be the first i between i_t and i_T for which $P(G_i|C)P(S_t|G_i, C)$ is maximized. Choose G_k as the solution.

Some basic results on the property of the enumerative Bayesian algorithm can be summarized as follows.

Theorem 7.7   The enumerative Bayesian algorithm determines a grammar with maximum a posteriori conditional probability $P(G|S_t, C)$ given the denumerable class C of stochastic grammars and the positive stochastic sample S_t.

Theorem 7.8   The enumerative Bayesian algorithm will determine at most one of a set of equivalent stochastic grammars.

Horning's approach to the inference of stochastic grammars provides a formal basis for choosing one stochastic grammar among many possible stochastic grammars. But due to the excessive computational cost of enumerating and testing all grammars in a class C, the proposed model for inference of stochastic grammar appears to present some difficulty in practical applications.

7.5   Synthesis of Stochastic Finite-State Automata

Following the approach taken by Biermann and Feldman [33], we shall now present a procedure for constructing a stochastic finite-state automaton from a finite set S_n of stochastic strings [38, 39]. The procedure used is to directly construct the states and state transitions of the stochastic finite-state automaton. The associated state transition probabilities can be determined from the probability information associated with the set of stochastic strings.

After some preliminary definitions and notations are established, the synthesis procedure of a stochastic finite-state automaton will be presented.

Definition 7.21 $S_n = \{(s_1, f_1), \ldots, (s_n, f_n)\}$, where $s_i \in V_T^*$, $\sum_{i=1}^n f_i = 1$, and the f_i's are rational numbers, is a finite set of stochastic strings. The characteristic set of S_n, denoted by $\bar{S}_n$, is $\{s_1, \ldots, s_n\}$. The set $\{f_1, \ldots, f_n\}$ is denoted by R_n, and f_i is often interpreted as the weight or probability associated with the string s_i.

Definition 7.22   Let S be a finite set of stochastic strings.

$$D_S = \{x \in V_T^* \mid xy \in \bar{S} \text{ for some } y \in V_T^*\} \tag{7.15}$$

Example 7.6   Let $S_3 = \{(a, \frac{1}{2}), (ab, \frac{1}{4}), (abb, \frac{1}{4})\}$. Then

$$\bar{S}_3 = \{a, ab, abb\}, \qquad R_3 = \{\tfrac{1}{2}, \tfrac{1}{4}, \tfrac{1}{4}\}, \qquad \text{and} \qquad D_S = \{\lambda, a, ab, abb\}$$

where λ is the empty string.

Definition 7.23   The k tail of $z \in D_S$ with respect to the set of stochastic strings S, denoted by $h(z, S, k)$, is defined as follows.

$$h(z, S, k) = \{(x, f) \,|\, (zx, f) \in S \text{ and } |x| \le k\} \qquad (7.16)$$

The function $h(z, S, k)$ is not defined for z not in D_S. The stochastic finite-state automaton that accepts the set S will be denoted by $S_a(S, k)$, where k is the maximal length of s in $\bar{S}$.

Definition 7.24   Let $S_a(S, k) = (Q, \Sigma, M, \pi_0, F)$ be a stochastic finite-state automaton accepting S where

(1) $k = \max_{s \in S}|s|$, $S = (\bar{S}, R)$ and $\bar{S} \in 2^{V_T^*}$
(2) $\Sigma = V_T$
(3) $Q = \{q_r\} \cup \{q \in 2^{V_T^* \times R} \,|\, h(z, S, k) = q \text{ for some } z \in D_S\} \qquad (7.17)$

$$Q' = \{q \in Q \,|\, (\lambda, f) \in q \quad \text{where} \quad f \in R\} \qquad (7.18)$$

$$F = \{q \in Q' \,|\, q \in \{(\lambda, f) \quad \text{for} \quad f \in R\}\} \qquad (7.19)$$

Now we merge all q in F into a single state q_f, the final state, and identify the set Q by $\{q_1, \ldots, q_{n-1}, q_n\}$, where $q_1 = h(\lambda, S, k)$, the initial state; $q_{n-1} = q_f$; and $q_n = q_r$, the rejection state.

(4)   Since q_1 is the initial state, $\pi_0 = [1, 0, \ldots, 0]$, an n-dimensional row vector.
(5)   Since q_{n-1} is designated as the final state, $\pi_F = [0, \ldots, 0, 1, 0]^T$, an n-dimensional column vector.
(6)   For each $a \in \Sigma$, the mapping $M(a) = [p_{ij}(a)]$ is defined as follows:

  (i) $p_{ij}(a) \ne 0$ if there is a $z \in \Sigma^*$ such that $h(z, S, k) = q_i$ and $h(za, S, k) = q_j$ for all $i, j = 1, \ldots, n - 2$.
  (ii) $p_{i, n-1}(a) \ne 0$ if there is a $q_j \in Q$ such that $p_{ij}(a) \ne 0$ for $i = 1, \ldots, n - 2$.
  (iii) $p_{n-1, j}(a) = 0$ for $j = 1, \ldots, n - 1$, and $p_{n-1, n}(a) = 1$.
  (iv) $p_{n, j}(a) = 0$ for $j = 1, \ldots, n - 1$ and $p_{n, n}(a) = 1$.

Furthermore, the $p_{ij}(a)$'s must satisfy the following normalization conditions.

  (a) $\displaystyle\sum_{a \in \Sigma} \sum_{j=1}^{n-1} p_{ij}(a) = 1$ for $i = 1, \ldots, n - 2$

  (b) $\displaystyle\sum_{j=1}^{n} p_{ij}(a) = 1$ for $i = 1, \ldots, n$ and $a \in \Sigma$.

Then by using the set of equations defined by

$$\pi_0 M(s_i)\pi_F = p(s_i)$$

for every $(s_i, p(s_i)) \in S$, all $p_{ij}(a)$'s can therefore be determined.

Since $L(S_a, 0) = \bar{S}$, and the $p_{ij}(a)$'s are determined such that

$$\pi_0 M(s_i)\pi_F = p(s_i) = f_i$$

for each $s_i \in \bar{S}$, $L(S_a) = S$.

Example 7.7　Let $S_3 = \{(a, 2/5), (ab, 2/5), (abb, 1/5)\}$. The stochastic finite-state automaton that accepts S_3 is defined as follows.

$$S_a(S_3, k) = (Q, \{a, b\}, M, \pi_0, F)$$

where

$$k = 3, \qquad Q = \{q_1, q_2, q_3, q_4, q_5\}$$

in which

$$q_1 = h(\lambda, S_3, 3) = \{(a, 2/5), (ab, 2/5), (abb, 1/5)\}$$
$$q_2 = h(a, S_3, 3) = \{(\lambda, 2/5), (bb, 2/5), (bb, 1/5)\}$$
$$q_3 = h(ab, S_3, 3) = \{(\lambda, 2/5), (b, 1/5)\}$$
$$q_4 = h(abb, S_3, 3) = \{(\lambda, 1/5)\}$$
$$q_5 = q_r. \qquad \text{(the rejection state)}$$

$$q_0 = q_1, \qquad q_f = q_4, \qquad \pi_0 = [1, 0, 0, 0, 0], \qquad \pi_F = [0, 0, 0, 1, 0]^T$$

and

$$M(a) = \begin{bmatrix} 0 & p_{12}(a) & 0 & p_{14}(a) & p_{15}(a) \\ 0 & 0 & 0 & 0 & 1 \\ 0 & 0 & 0 & 0 & 1 \\ 0 & 0 & 0 & 0 & 1 \\ 0 & 0 & 0 & 0 & 1 \end{bmatrix}$$

$$M(b) = \begin{bmatrix} 0 & 0 & 0 & 0 & 1 \\ 0 & 0 & p_{23}(b) & p_{24}(b) & p_{25}(b) \\ 0 & 0 & 0 & p_{34}(b) & p_{35}(b) \\ 0 & 0 & 0 & 0 & 1 \\ 0 & 0 & 0 & 0 & 1 \end{bmatrix}$$

where

$$p_{12}(a) + p_{14}(a) + p_{15}(a) = 1, \qquad p_{12}(a) + p_{14}(a) = 1$$
$$p_{23}(b) + p_{24}(b) + p_{25}(b) = 1, \qquad p_{23}(b) + p_{24}(b) = 1$$
$$p_{34}(b) + p_{35}(b) = 1, \qquad p_{34}(b) = 1$$

It can be shown that

$$p_{15}(a) = p_{25}(b) = p_{35}(b) = 0$$

Since

$$\pi_0 M(a)\pi_F = p_{14}(a) = 2/5$$

we have

$$p_{12}(a) = 1 - p_{14}(a) = 3/5$$

Since

$$\pi_0 \, M(ab)\pi_F = p_{12}(a)p_{24}(b) = 2/5$$

we have

$$p_{24} = \tfrac{2}{3} \quad \text{and} \quad p_{23}(b) = 1 - p_{24}(b) = \tfrac{1}{3}$$

It can be easily shown that $\pi_0 \, M(abb)\pi_F = 1/5$.

After the stochastic automaton is synthesized, a corresponding stochastic finite-state grammar can be constructed using a procedure similar to that used in the nonstochastic case (Chapter 2). Associated with the initial state, the start symbol S is assigned. For the remaining states in Q, assign a nonterminal for each of the states. Then, with the set of terminals $V_T = \Sigma$,

$$S \xrightarrow{p} aA \qquad \text{if } (q_0, a) = (q, p), \ a \in \Sigma, \text{ where } A \text{ is the nonterminal associated with the state } q \text{ and } q \notin F;$$

$$A_1 \xrightarrow{p} aA_2 \qquad \text{if } (q_1, a) = (q_2, p), \ a \in \Sigma \text{ where } A_1 \text{ and } A_2 \text{ are the nonterminals associated with the states } q_1 \text{ and } q_2, \text{ respectively, and } q_2 \notin F;$$

$$A \xrightarrow{p} b \qquad \text{if } (q, b) = (q_f, p), \ b \in \Sigma \text{ and } q_f \in F, \text{ where } A \text{ is the nonterminal associated with the state } q.$$

Based on the construction procedure sketched, the stochastic finite-state grammar corresponding to the stochastic finite automaton is $G_s = (V_N, V_T, P_s, S)$, where $V_N = Q - \{q_f \cup q_r\} = \{q_1, q_2, q_3\}$, $V_T = \Sigma = \{a, b\}$, $S = q_1$, and P_s:

$$q_1 \xrightarrow{3/5} aq_2 \qquad q_2 \xrightarrow{2/3} b$$

$$q_1 \xrightarrow{2/5} a \qquad q_3 \xrightarrow{1} b$$

$$q_2 \xrightarrow{1/3} bq_3$$

Note that the language generated by G_s is precisely the given set of stochastic strings S_3.

Example 7.8   The stochastic finite-state grammar G_1 in Example 6.1 can be inferred from the sample

$$S_9 = \{(ab_1 c_1, 1/36), (ab_1 c_2, 2/36), (ab_1 c_3, 3/36),$$
$$(ab_2 c_1, 1/36), (ab_2 c_2, 21/36), (ab_2 c_3, 2/36),$$
$$(ab_3 c_1, 3/36), (ab_3 c_2, 2/36), (ab_3 c_3, 1/36)\}$$

Let $S_a(S_9, k) = (Q, \{a, b_1, b_2, b_3, c_1, c_2, c_3\}, M, q_0, F)$. For $k = 3$,

$$q_0 = h(\lambda, S_9, 3) = \{(ab_1c_1, 1/36), (ab_1c_2, 2/36), (ab_1 c_3, 3/36),$$
$$(ab_2 c_1, 1/36), (ab_2 c_2, 21/36), (ab_2 c_3, 2/36)$$
$$(ab_3 c_1, 3/36), (ab_3 c_2, 2/36), (ab_3 c_3, 1/36)\}$$

$$q_1 = h(a, S_9, 3) = \{(b_1c_1, 1/36), (b_1c_2, 2/36), (b_1c_3, 3/36).$$
$$(b_2 c_1, 1/26), (b_2 c_2, 21/36), (b_2 c_3, 2/36),$$
$$(b_3 c_1, 3/36), (b_3 c_2, 2/36), (b_3 c_3, 1/36)\}$$

$$q_2 = h(ab_1, S_9, 3) = \{(c_1, 1/36), (c_2, 2/36), (c_3, 3/36)\}$$

$$q_3 = h(ab_2, S_9, 3) = \{(c_1, 1/36), (c_2, 21/36), (c_3, 2/36)\}$$

$$q_4 = h(ab_3, S_9, 3) = \{(c_1, 3/36), (c_2, 2/36), (c_3, 1/36)\}$$

$$q_5 = h(ab_1c_1, S_9, 3) = \{(\lambda, 1/36)\}$$

$$q_6 = h(ab_1c_2, S_9, 3) = \{(\lambda, 2/36)\}$$

$$q_7 = h(ab_1c_3, S_9, 3) = \{(\lambda, 3/36)\}$$

$$q_8 = h(ab_2 c_2, S_9, 3) = \{(\lambda, 21/36)\}$$

$$q_9 = h(ab_2 c_1, S_9, 3) = q_5$$

$$q_{10} = h(ab_2 c_3, S_9, 3) = q_6$$

$$q_{11} = h(ab_3 c_1, S_9, 3) = q_7$$

$$q_{12} = h(ab_3 c_2, S_9, 3) = q_6$$

$$q_{13} = h(ab_3 c_3, S_9, 3) = q_5$$

Thus,

$$Q = \{q_0, q_1, q_2, q_3, q_4, q_5, q_6, q_7, q_8\}$$
$$F = \{(\lambda, 1/36), (\lambda, 2/36), (\lambda, 3/36), (\lambda, 21/36)\}$$
$$Q' = \{q_5, q_6, q_7, q_8\}$$

and

$$\delta(q_0, a) = (q_1, p_1), \qquad \delta(q_3, c_1) = (q_5, p_8)$$
$$\delta(q_1, b_1) = (q_2, p_2), \qquad \delta(q_3, c_2) = (q_8, p_9)$$
$$\delta(q_1, b_2) = (q_3, p_3), \qquad \delta(q_3, c_3) = (q_6, p_{10})$$
$$\delta(q_1, b_3) = (q_4, p_4), \qquad \delta(q_4, c_1) = (q_7, p_{11})$$
$$\delta(q_2, c_1) = (q_5, p_5), \qquad \delta(q_4, c_2) = (q_6, p_{12})$$
$$\delta(q_2, c_2) = (q_6, p_6), \qquad \delta(q_4, c_3) = (q_5, p_{13})$$
$$\delta(q_2, c_3) = (q_7, p_7),$$

The state transition diagram of the stochastic automaton $S_a(S_9, 3)$ is given in Fig. 7.4.

Using the normalization conditions and the information of S_9 (see Example 6.1), we obtain

$$p_1 = 1, \quad p_2 = 1/6, \quad p_3 = 2/3, \quad p_4 = 1/6, \quad p_5 = 1/6, \quad p_6 = 1/3, \quad p_7 = 1/2,$$
$$p_8 = 1/24, \quad p_9 = 21/24, \quad p_{10} = 1/12, \quad p_{11} = 1/2, \quad p_{12} = 1/3, \quad p_{13} = 1/6$$

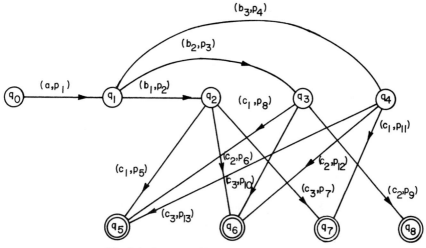

Fig. 7.4. State transition diagram of $S_a(S_9, 3)$.

The corresponding stochastic finite-state grammar can then be easily constructed from the synthesized stochastic automaton.

It should be noted that in this example $\sum_{i=1}^{9} f_i = 1$. The value of $k \, (=3)$ has been selected such that the resulting automaton will accept exactly the strings in S_9. If, for example, the information regarding the last two strings, $(ab_3c_2, 2/36)$, $(ab_3c_3, 1/36)$, is not available, then the transition from q_4 is only partially specified (i.e., only the transition from q_4 to q_7 is specified). In such a situation, the transition from q_4 to other states (including itself) except q_7 can be rather arbitrarily specified as long as the normalization conditions are satisfied. Of course, the arbitrarily specified transitions would affect the additional strings (and their associated probabilities) accepted by the automaton.†

7.6   A Practical Grammatical Inference System

Lee and Fu have proposed a practical system for grammatical inference [40]. The system utilizes the man–machine interactive capability to simplify the inference problem. A block diagram of the system is shown in Fig. 7.5. A human trainer, perhaps through an interactive graphic display, divides each string in the sample S_t into n different substrings. The sample set S_t is then divided into n substring sets. Each substring set represents, say, a set of sub-patterns with simple structures which can be easily described by a simple

---

† This might be interpreted as the generalization or the inductive part of the inference (learning) process.

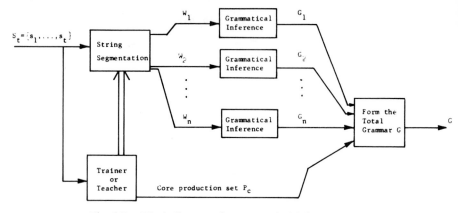

Fig. 7.5.   Block diagram of a grammatical inference system.

grammar, a grammar of the type that can be efficiently inferred by existing techniques. For pictorial (line) patterns, the following guidelines are suggested for the segmentation of a pattern into subpatterns:

(1)   Select subpatterns with simply connected structures, so the grammars for each substring set (representing subpatterns) can be easily inferred.

(2)   Segment the pattern so that full advantage can be taken from repetitive subpatterns. This segmentation allows the use of one prototype grammar which accepts a frequently encountered substring, and also results in supplying more sample substrings for inference.

After the sample string set S_t is divided into n substring sets $W_1, W_2, \ldots, W_n$, we can infer n distinct grammars $G_1, \ldots, G_n$, one for each substring set. Let the grammar inferred from W_j be

$$G_j = (V_{Nj}, V_{Tj}, P_j, S_j), \qquad j = 1, \ldots, n,$$

and the total grammar (the grammar describing the total patterns) be

$$G = (V_N, V_T, P, S)$$

The terminal set V_{Tj} belongs to V_T with the possible exception of a set of connecting symbols C_j. Symbols C_j are treated as terminals in G_j and as nonterminals in G, that is, $C_j \in V_{Tj}$ and $C_j \in V_N$. The following examples illustrate the procedure.

Example 7.9   Consider the set of noisy right-angle triangles shown in Fig. 7.6. Preprocessing has been performed, so the pattern fits into an 8×8 picture matrix, and it is chain coded by the following primitive set.

$$V_T = \left\{ \underset{-3}{\diagup}, \underset{-2}{\leftarrow}, \underset{-1}{\diagdown}, \underset{0}{\uparrow}, \underset{+1}{\diagup}, \underset{+2}{\rightarrow}, \underset{+3}{\diagdown}, \underset{+4}{\diagdown} \right\}$$

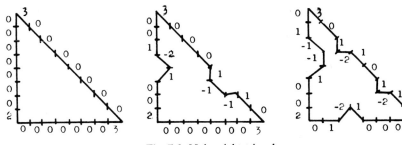

Fig. 7.6. Noisy right triangle.

Let the grammar inferred be $G = (V_N, V_T, P, S)$. The core productions are given as

$$S \to RBDB$$

which represents the two sides and the right angle;

$$D \to 3H3$$

which represents the acute angle and hypotenuse; and

$$R \to 020$$

which represents the right angle. These core productions are not unique. They are chosen this way just for convenience of representation.† There are two grammars, G_1 and G_2, to be inferred.

$$G_1 = (V_{N1}, V_{T1}, P_1, H) \qquad \text{for hypotenuse}$$
$$G_2 = (V_{N2}, V_{T2}, P_2, B) \qquad \text{for side}$$

where $H \in V_{N1}$ and $B \in V_{N2}$.

The Crespi-Reghizzi inference algorithm for k-distinct and k-homogeneous grammars is used to infer the two grammars. We first form two sets of substrings, W_1 and W_2. W_1 represents all substring configurations of the "hypotenuse" subpattern, and W_2 the "side" subpattern. All the strings in W_1 and W_2 must be structured. For the three sample patterns in Fig. 7.6, we obtain

$$W_1 = \{[0[0[0[0[0[0[0[0]]]]]]]], [0[0[0[1 \quad -1 \quad -1 \quad 1]0]]], [0[1 \quad -2 \quad 1]0$$
$$[1 \quad -2 \quad 1]0]\}$$

$$W_2 = \{[0[0[0[0[0[0[0]]]]]]], [0[0[0[0[0[0[0]]]]]]], [0[0[0[0[0[0[0]]]]]]],$$
$$[0[[1 \quad -2 \quad 1]0[0]]], [0[1 \quad -1 \quad -1 \quad 1]0], [0[0[0[1 \quad -2 \quad 1]]]]\}$$

---

† For example, an alternative selection of core productions is to use a single production $S \to 020B3H3B$.

The inferred grammar G_1 from W_1 is $G_1 = (V_{N1}, V_T, P_1, H)$, where $V_{N1} = \{H, M_1, M_2, M_3\}$ and P_1:

$$H \to M_1, \qquad H \to M_2$$
$$M_1 \to 0, \qquad M_1 \to 0M_30, \qquad M_1 \to 0M_30M_30$$
$$M_2 \to 0M_1, \qquad M_2 \to 0M_2$$
$$M_3 \to 1 \quad -1 \quad -1 \quad 1, \qquad\qquad M_3 \to 1 \quad -2 \quad 1$$

Similarly, $G_2 = (V_{N2}, V_T, P_2, B)$, where $V_{N2} = \{B, N_1, N_2, N_3, N_4\}$ and P_2:

$$B \to N_1, \qquad B \to N_2$$
$$N_1 \to 0, \qquad N_1 \to 0N_30$$
$$N_2 \to 0N_1, \qquad N_2 \to 0N_2, \qquad N_2 \to 0N_3, \qquad N_2 \to 0N_4$$
$$N_3 \to 1 \quad -1 \quad -1 \quad 1 \qquad N_3 \to 1 \quad -2 \quad 1$$
$$N_4 \to N_30N_1$$

Combining the two grammars G_1 and G_2 with the core productions, we get

$$G = (V_N, V_T, P, S)$$

where

$$V_N = \{S, R, B, D, H, M_1, M_2, M_3, N_1, N_2, N_3, N_4\}$$
$$V_T = \{0, 1, -1, 2, -2, 3, -3, 4\}$$

and P:

$$S \to RBDB, \qquad D \to 3H3, \qquad R \to 020$$
$$B \to N_1, \qquad B \to N_2, \qquad H \to M_1, \qquad H \to M_2$$
$$M_1 \to 0, \qquad M_1 \to 0M_30, \qquad M_1 \to 0M_30M30$$
$$M_2 \to 0M_1, \qquad M_2 \to 0M_2,$$
$$M_3 \to 1 \quad -1 \quad -1 \quad 1, \qquad M_3 \to 1 \quad -2 \quad 1$$
$$N_1 \to 0N_30, \qquad N_1 \to 0$$
$$N_2 \to 0N_1, \qquad N_2 \to 0N_2, \qquad N_2 \to 0N_3, \qquad N_2 \to 0N_4$$
$$N_3 \to 1 \quad -1 \quad -1 \quad 1, \qquad N_3 \to 1 \quad -2 \quad 1$$
$$N_4 \to N_30N_1$$

This grammar will generate the three right-angle triangles and other variations of noisy triangles.

Instead of applying Crespi-Reghizzi's inference algorithm, we certainly can also use Biermann–Feldman's inference procedure for finite-state grammars to infer G_1 and G_2.

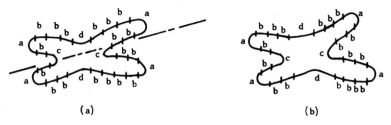

Fig. 7.7. Submedian chromosome sample. (a) Coded chromosome, where $x =$ *cdabbbdbbbabbcbbabbbbdbbabb*. The chromosome has the same structure above and below the dashed line. (b) Substructures of coded chromosomes. Structured samples of S_1, the set of all arm strings $S_1 = \{[b[[[a]b]b]b]; [b[b[b[a]]b]b]; [b[b[[[a]b]b]b]b]; [b[b[a]]b]\}$.

Example 7.10   Consider the submedian chromosome pattern shown in Fig. 7.7a. Making use of the symmetry of the pattern, we can choose the subpatterns as shown in Fig. 7.7b. The core productions are

$$S \to BB$$
$$B \to c \text{ Arm } d \text{ Arm}$$

where B represents the subpattern of half the chromosome. The subpattern that needs to be inferred in this case is "Arm."

Crespi-Reghizzi's inference algorithm is again used here. The structured string sample set for Arm is (Fig. 7.7b)

$$\{[b[[[a]b]b]b], [b[b[b[a]]b]b]; [b[b[[[a]b]b]b]b]; [b[b[a]]b]\}$$

These four substrings are then used to infer the grammar for generating Arm. The inferred grammar is

$$G_1 = (V_{N1}, V_T, P_1, \text{Arm})$$

where

$$V_{N1} = \{\text{Arm}, H, G, F, E\}$$
$$V_T = \{ \widehat{} a, \!\!\not\!/ b, \,\rangle c, \rightleftharpoons d \}$$

and P_1:

$$
\begin{array}{ll}
\text{Arm} \to G & G \to bGb \\
G \to bFb & G \to bHb \\
F \to Eb & F \to Fb \\
H \to bE & E \to a
\end{array}
$$

Thus, the grammar for the submedian chromosome patterns is $G = (V_N, V_T,$

$P, S)$, where $V_N = \{S, B, \text{Arm}, H, G, F, E\}$ and P:

$$S \to BB$$
$$B \to c \,\text{Arm}\, d \,\text{Arm}$$
$$\text{Arm} \to G$$

$$G \to bGb \qquad G \to bFb \qquad G \to bHb$$
$$F \to Fb \qquad F \to Eb$$
$$H \to bE \qquad E \to a$$

7.7   Approximation of Stochastic Languages

The complexity of a class of languages is generally defined in terms of its syntactic structure. For the phrase-structure grammars, unrestricted, context-sensitive, context-free, and finite-state languages are arranged in descending order of their complexity. That is, unrestricted languages are considered to be the most complex languages, while finite-state languages are the least complex ones.

Given a language L, it is known that the more complex the L, the harder it is to infer a grammar that generates L. The main reason is that the existing grammatical inference algorithms require that every string defined in L must be generated by the inferred grammar G, while none of the other strings not in L can be generated by G. But for a stochastic language (L, p), we can associate with each string x in L a probability of generation $p(x)$. It is reasonable to assume that the string which is most likely to be generated is the most significant string in L. The least significant string in L is the one that is least likely to be generated. Therefore, a possible realistic approach for inferring a grammar of a stochastic language (L, p) is to find one which will generate those strings with relatively high probability of occurrence and may or may not generate the other strings in L. This concept is particularly useful for applications to syntactic pattern recognition because we can always construct a relatively simple stochastic finite-state automaton to accept those strings with a relatively high probability of occurrence. Before we present this approach to inferring a grammatical structure of a stochastic language, we shall first introduce some basic definitions and results of the approximation of stochastic language by a stochastic finite-state language.

Definition 7.25   Let $\bar{S} \in 2^{V_T^*}$, where V_T is a finite set of (terminal) symbols. The probability of the set S is

$$p(\bar{S}) = \sum_{x \in S} p(x) \qquad (7.20)$$

where $p(x)$ is a mapping of x in V_T^* to $[0, 1]$ and $\sum_{x \in V_T^*} p(x) = 1$.

Definition 7.26 Let $0 < \varepsilon < 1$. A stochastic language (L_2, p_2) weakly ε-approximates a stochastic language (L_1, p_1) if

$$p_i(L_1 \cap L_2) \geq 1 - \varepsilon \quad \text{for} \quad i = 1, 2 \quad (7.21)$$

where

$$p_i(L_i \cap L_2) = \sum_{x \in L_1 \cap L_2} p_i(x) \quad \text{for} \quad i = 1, 2 \quad (7.22)$$

Definition 7.27 Let $0 < \varepsilon < 1$. A stochastic language (L_2, p_2) strongly ε-approximates a stochastic language (L_1, p_1) if

$$\text{(i)} \quad p_i(L_1 \cap L_2) > 1 - \varepsilon \quad \text{for} \quad i = 1, 2 \quad (7.23)$$

and

$$\text{(ii)} \quad p_1(x) = p_2(x) \quad \text{for all } x \text{ in } L_1 \cap L_2 \quad (7.24)$$

Example 7.11 Consider a stochastic context-sensitive language (L_1, p_1) where $L_1 = \{a^n b^n c^n \,|\, n > 0\}$ and $p_1(a^n b^n c^n) = (1/2)^n$, $n > 0$. The language can be generated by the grammar G_1.

$$G_1 = (V_N, V_T, P_s, S)$$

where $V_N = \{S, A, B\}$, $V_T = \{a, b, c\}$, and P_s :

$$S \xrightarrow{1/2} aSBA \qquad bB \xrightarrow{1} bb$$

$$S \xrightarrow{1/2} aBA \qquad bA \xrightarrow{1} bc$$

$$AB \xrightarrow{1} BA \qquad cA \xrightarrow{1} cc$$

$$aB \xrightarrow{1} ab$$

The following stochastic context-free grammar G_2 will generate a stochastic context-free language (L_2, p_2) to approximate at least 96.9% of the strings in L_1 [41]. In other words, (L_2, p_2) (weakly) ε-approximates (L_1, p_1) with $\varepsilon \leq 0.03125$.

$$G_2 = (V_N, V_T, P_s, S)\dagger$$

† The grammar G_2 can be inferred by using first the procedure described by Crespi-Reghizzi [31] and then the probability inference algorithm introduced by Lee and Fu [40]. This two-step inference scheme for stochastic context-free grammars has also been applied to the chromosome classification problem [40, 41].

where $V_N = \{S, A, B, C, E, F, H, J, K\}$, $V_T = \{a, b, c\}$, and P_s:

$$S \xrightarrow{1} aAc \qquad D \xrightarrow{0.492} aEc \qquad H \xrightarrow{1} bb$$

$$A \xrightarrow{0.502} b \qquad E \xrightarrow{0.516} HH \qquad J \xrightarrow{0.365} aJc$$

$$A \xrightarrow{0.498} aBc \qquad E \xrightarrow{0.484} aFc \qquad J \xrightarrow{0.635} KHHb$$

$$B \xrightarrow{0.504} bb \qquad F \xrightarrow{0.538} HHb \qquad K \xrightarrow{0.365} bK$$

$$B \xrightarrow{0.496} aDc \qquad F \xrightarrow{0.462} aJc \qquad K \xrightarrow{0.635} b$$

$$D \xrightarrow{0.508} Hb$$

$$L(G_2) = (L_2, p_2) = \{(a^i b^i c^i, p_i) \,|\, i = 1, 2, 3, 4, 5 \text{ and } (a^n b^m c^n, p_{nm}) \,|\, n, m \geq 6\}$$

The probabilities of some of the strings generated by G_1 and G_2, respectively, are displayed in the following table.

Strings x	$p_1(x)$	$p_2(x)$
abc	0.5	0.502
$a^2 b^2 c^2$	0.25	0.251
$a^3 b^3 c^3$	0.125	0.1257
$a^4 b^4 c^4$	0.0625	0.0629
$a^5 b^5 c^5$	0.03125	0.0317

Theorem 7.9 Any stochastic context-free or finite-state language can be strongly ε-approximated by a stochastic finite-state language, for some $0 < \varepsilon < 1$.†

Proof Let (L, p) be a stochastic context-free or finite-state language where $L = \{x_1, x_2, \ldots\}$ and $p(x_1) \geq p(x_2) \cdots$. Since $\sum_{i=1}^{\infty} p(x_i) = 1$, there exists an $n < \infty$ [45] such that

$$\sum_{i=n+1}^{\infty} p(x_i) < \varepsilon \tag{7.25}$$

where $\varepsilon > 0$. Therefore, let (L', p') be some stochastic language such that

$$L' = \{x_1, \ldots, x_n\} \cup \{y\}, \qquad p'(x_i) = p(x_i) \qquad \text{for} \quad i = 1, \ldots, n$$

† It was pointed out to the author recently by Professor L. A. Zadeh that it should also be interesting to approximate less complex languages (e.g., finite state or context free) by more complex languages (e.g., context free or context sensitive) with the intention of reducing significantly the number of productions.

and

$$p'(y) = \sum_{i=n+1}^{\infty} p(x_i) \qquad (7.26)$$

where y is a string not in $\{x_1, \ldots, x_n\}$.

Clearly, (L', p') is a stochastic finite-state language that strongly ε-approximates the stochastic language (L, p).

Example 7.12   Consider a stochastic context-free language (L, p) where $L = \{0^n1^n \mid n > 0\}$ and $p(0^n1^n) = (\frac{1}{2})^n$ for $n > 0$. For any $0 < \varepsilon < 1$, a stochastic finite-state language (L', p') which strongly ε-approximates (L, p) can be defined as follows. Let $L' = \{0^n1^n \mid 0 < n \le N\} \cup \{x\}$, where $x \in \{0, 1\}^* - L$; and $p'(0^n1^n) = p(0^n1^n)$ for $0 < n \le N$ and $p'(x) = 1 - \sum_{n=1}^{N} p(0^n1^n) = (\frac{1}{2})^N$. It is easy to show that

$$p(L \cap L') = p'(L \cap L') = \sum_{n=1}^{N} \left(\frac{1}{2}\right)^n = 1 - \left(\frac{1}{2}\right)^N$$

If we choose N such that $(\frac{1}{2})^N < \varepsilon$,

$$p(L \cap L') = p'(L \cap L') > 1 - \varepsilon$$

Hence, (L', p') strongly ε-approximates (L, p).

Corollary   Any stochastic context-free or finite-state language can be weakly ε-approximated by a stochastic finite-state language.

Based on the concept of ε-approximation of a stochastic language and the method of constructing a stochastic finite-state automaton (Section 7.5) that accepts a stochastic finite-state language, we shall now describe an approach to the inference of stochastic languages.

The inference model for stochastic languages consists of three basic parts.

(1)   A stochastic syntactic source G from which strings are statistically generated for observation.

(2)   An inference algorithm which determines a stochastic finite-state grammar G_t which will generate a stochastic language to ε-approximate the stochastic language generated by the stochastic syntactic source G based on t observations of $L(G)$. If necessary, a stochastic context-free or context-sensitive language can then be constructed to ε-approximate the inferred stochastic finite-state language.

(3)   An informant who has information about the structure of the stochastic syntactic source is used to evaluate the performance of the inference machine. That is, for some given $0 < \varepsilon < 1$, if $p(L(G) \cap L(G_t)) \ge 1 - \varepsilon$, where G_t is the inferred grammar based on t observations of $L(G)$, G_t is an

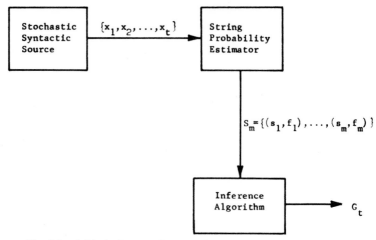

Fig. 7.8   A block diagram for the inference of stochastic grammars.

acceptable solution; otherwise, $(t + 1)$ observations are needed to infer the syntactic structure of the stochastic source.

The inference system for a stochastic language is shown schematically in Fig. 7.8. Let $x_1, \ldots, x_t$ be a set of t observations in which there are $n_1\, s_1$, $n_2\, s_2, \ldots$, and $n_m\, s_m$, where $n_1 + n_2 + \cdots + n_m = t$ and $s_1, \ldots$, and s_m are distinctly different strings. The probability of generation associated with each string s_i is empirically estimated by the relative frequency of occurrence, that is, $f_i = n_i/t$. Then the set of stochastic strings S_m presented to the inference machine is $\{(s_1, f_1), \ldots, (s_m, f_m)\}$, where

$$\sum_{i=1}^{m} f_i = \sum_{i=1}^{m} n_i/t = 1$$

The stochastic finite-state automaton that accepts S_m is defined as $S_a(S_m, k)$ where k is the maximal length of strings in S_m.

This inference scheme provides a stochastic finite-state grammar which will generate a stochastic language to ε-approximate the stochastic language generated by the unknown stochastic syntactic source. The inferred stochastic finite-state grammar may predict other strings that can be generated by the unknown stochastic syntactic source as well as some that can not be generated by the source. But it provides a probabilistic description of the observed strings generated by the unknown stochastic syntactic source G. As t approaches infinity, the relative frequency f_i of occurrence of each observed string s_i approaches the true probability of generation $p(s_i)$ defined by the unknown stochastic syntactic source G. Consequently, $p(L(G) \cap L(G_t))$ approaches 1 as t approaches ∞.

Recently, Patel [42] proposed an inference algorithm for stochastic finite-state languages in terms of growing grammars (grammars with an increasing number of productions inferred at each step of the algorithm). A growing grammar becomes a solution when a point is reached where it is completely defined as an unambiguous right-linear stochastic grammar that is proper and consistent [45,46]. The exact solution is obtained if all the rules of the grammar being inferred are used at least once in generating the given sample. Otherwise, an approximation of the language inferred will be obtained.

References

1. N. Chomsky, *Aspects of the Theory of Syntax*. MIT Press, Cambridge, Massachusetts, 1964.
2. P. Naur, *et al.*, Report on the algorithmic language ALGOL 60. *Comm. ACM* **3**, 299, 314 (1960).
3. E. T. Irons, A syntax-directed compiler for ALGOL 60. *Comm. ACM* **4**, 51–55 (1961).
4. B. M. Leavenworth, FORTRAN IV as a syntax language. *Comm. ACM* **7**, 72–80 (1964).
5. R. W. Floyd, The syntax of programming languages—A survey. *IEEE Trans. Electron. Comput.* **EC-13**, 346–353 (1964).
6. K. S. Fu and P. H. Swain, On syntactic pattern recognition. In *Software Engineering* (J. T. Tou, ed.), Vol. 2, Academic Press, New York, 1971.
7. R. Sherman and G. W. Ernst, Learning patterns in terms of other patterns. *Pattern Recognition* **1**, 301–314 (1969).
8. R. Sauvain and L. Uhr, A teachable pattern describing and recognizing program. *Pattern Recognition* **1**, 219–232 (1969).
9. P. H. Whinston, *Learning Structural Descriptions from Examples*. Rep. MAC TR-76, Project MAC. MIT, Cambridge, Massachusetts, September 1970.
10. S. Kaneff, On the role of learning in picture processing. In *Picture Language Machines* (S. Kaneff, ed.). Academic Press, New York, 1970.
11. I. D. G. Macleod, On finding structure in pictures. In *Picture Language Machines* (S. Kaneff, ed.). Academic Press, New York, 1970.
12. J. C. Schwebel, A graph structure model for structure inference. In *Graphics Languages* (F. Nake and A. Rosenfeld, eds.). North-Holland Publ., Amsterdam, 1972.
13. D. B. Peizer and D. L. Olmsted, Modules of grammar acquisition. *Language* **45**, 61–96 (1969).
14. R. J. Solomonoff, A new method for discovering the grammars of phrase structure languages. *Information Processing*. UNESCO Publ. House, New York, 1959.
15. S. Crespi-Reghizzi, M. A. Melkanoff, and L. Lichten, *A Proposal for the Use of Grammar Inference as a Tool for Designing Programming Languages*. Rep. No. 71-1. Lab. di Calcolatori, Inst. di Elettrotecn. ed Electtron. del Politecn. Milano, Italy, March 1971.
16. J. A. Feldman and D. Gries, Translator writing systems. *Comm. ACM* **11**, 77–113 (1968).
17. S. Kaneff, ed., *Picture Language Machines*. Academic Press, New York, 1970.
18. D. T. Ross and J. E. Rodriguez, Theoretical foundations for the computer-aided design project. *Proc. AFIPS Spring Joint Comput. Conf., April 1963*. Spartan Books, Inc., Baltimore, Maryland, 1963.

19. P. S. Rosenbaum, A grammar base question-answering procedure. *Comm. ACM* **10**, 630–635 (1967).
20. T. G. Evans, Grammatical inference techniques in pattern analysis. In *Software Engineering* (J. T. Tou, ed.), Vol. 2. Academic Press, New York, 1971.
21. J. A. Feldman and P. Shields. *On Total Complexity and the Existence of Best Programs.* Tech. Rep. No. CS-255. Comput. Sci. Dept., Stanford Univ., Stanford, California, 1972.
22. E. M. Gold, Language identification in the limit. *Information and Control* **10**, 447–474 (1967).
23. J. A. Feldman, *First Thought on Grammatical Inference.* Stanford Artificial Intelligence Proj. Memo. No. 55. Stanford Univ., Stanford, California, 1967.
24. J. A. Feldman, *Some Decidability Results on Grammatical Inference and Complexity.* Stanford Artificial Intelligence Proj. Memo. AI-93. Stanford Univ., Stanford, California, 1969.
25. J. A. Feldman, J. Gips, J. J. Horning, and S. Reder, *Grammatical Complexity and Inference.* Tech. Rep. No. CS-125. Comput. Sci. Dept., Stanford Univ., Stanford, California, 1969.
26. J. J. Horning, *A Study of Grammatical Inference.* Tech. Rep. No. CS-139. Comput. Sci. Dept., Stanford Univ., Stanford, California, 1969.
27. J. J. Horning, A procedure for grammatical inference. *IFIP Congr., Yugoslavia, August 1971.*
28. T. W. Pao, *A Solution of the Syntactical Induction-Inference Problem for a Non-Trivial Subset of Context-Free Languages,* Interim Tech. Rep. No. 69-19. Moore School of Elec. Eng., Univ. of Pennsylvania, Philadelphia, Pennsylvania, 1969.
29. S. Crespi-Reghizzi, *The Mechanical Acquisition of Precedence Grammars.* Tech. Rep. UCLA-Eng-7054. Univ. of California, 1970.
30. S. Crespi-Reghizzi, An effective model for grammar inference. *IFIP Congr., Yugoslavia, August 1971.*
31. S. Crespi-Reghizzi, Reduction of enumeration in grammar acquisition. *Proc. Int. Joint Conf. Artificial Intelligence, 2nd, London, September 1–3, 1971.*
32. R. J. Solomonoff, A formal theory of inductive inference. *Information and Control* **7**, 1–22, 224–254 (1964).
33. A. W. Biermann and J. A. Feldman, *On the Synthesis of Finite-State Acceptors.* Stanford Artificial Intelligence Proj. Memo. AIM-114. Stanford Univ., Stanford, California, 1970.
34. A. W. Biermann, A Grammatical inference program for linear languages. *Int. Conf. Syst. Sci., 4th, Hawaii, January 1971.*
35. A. W. Biermann and J. A. Feldman, A survey of results in grammatical inference. *Int. Conf. Frontiers of Pattern Recognition, Honolulu, Hawaii, January 18–20, 1971.*
36. A. W. Biermann and J. A. Feldman, On the synthesis of finite-state machines from samples of their behavior. *IEEE Trans. Computers* **C-21**, 592–597 (1972).
37. J. A. Feldman, Some decidability results on grammatical inference and complexity. *Information and Control* **20**, 244–262 (1972).
38. K. S. Fu, On syntactic pattern recognition and stochastic languages. *Int. Conf. Frontiers of Pattern Recognition, Honolulu, Hawaii, January, 18–20, 1971;* published in *Frontiers of Pattern Recognition* (S. Watanabe, ed.). Academic Press, New York, 1972.
39. T. Huang and K. S. Fu, *Stochastic Syntactic Analysis and Syntactic Pattern Recognition.* Tech. Rep. TR-EE 72-5. School of Elec. Eng., Purdue Univ., Lafayette, Indiana, 1972.
40. H. C. Lee and K. S. Fu, A syntactic pattern recognition system with learning capability. *Proc. Int. Symp. Comput. and Inform. Sci. (COINS-72), Miami Beach, Florida, December 14–16, 1972.*

41. H. C. Lee and K. S. Fu, *Stochastic Linguistics for Pattern Recognition.* Tech. Rep. TR-EE 72-17. School of Elec. Eng., Purdue Univ., Lafayette, Indiana, June 1972.

42. A. R. Patel, *Grammatical Inference for Probabilistic Finite State Languages.* Ph.D. Thesis, Dept. of Elec. Eng., Univ. of Connecticut, Storrs, Connecticut, May 1972.

43. J. C. Reynolds, *Grammatical Covering.* Tech. Memo. No. 96. Appl. Math. Div. Argonne Nat. Lab. Argonne, Illinois, March 1968.

44. J. Hartmanis and R. E. Stearns, *Algebraic Structure Theory of Sequential Machines.* Prentice-Hall, Englewood Cliffs, New Jersey, 1966.

45. T. L. Booth, Probability representation of formal languages. *IEEE Annu. Symp. Switching and Automata Theory, 10th, October 1969.*

46. K. S. Fu and T. Huang, Stochastic grammars and languages. *Int. J. Comput. Inform. Sci.* **1**, 135–170 (1972).

47. S. Watanabe, *Knowing and Guessing*, Wiley, New York, 1969.

48. E. M. Gold, Current approaches to the inference of phrase structure grammars. *Phrase Struct. Miniconf. Montreal, Canada, March 22–23, 1973.*

49. R. H. Anderson, The linguistic approach to pattern recognition as a basis for adaptive program behavior. *Proc. Allerton Conf. Circuit. and Syst. Theory, Monticello, Illinois, 1972*, pp. 2–10.

Appendix A

Syntactic Recognition of Chromosome Patterns

Ledley *et al.* [1] have proposed the following context-free grammar for the description of submedian and telocentric chromosome patterns (Fig. A.1).

$G = (V_N, V_T, P, \{\langle\text{submedian chromosome}\rangle, \langle\text{telocentric chromosome}\rangle\})$

where

$V_N = \{\langle\text{submedian chromosome}\rangle, \langle\text{telocentric chromosome}\rangle, \langle\text{arm pair}\rangle,$
$\langle\text{left part}\rangle, \langle\text{right part}\rangle, \langle\text{arm}\rangle, \langle\text{side}\rangle, \langle\text{bottom}\rangle\}$

$V_T = \left\{ \overset{\cdot}{\frown}, \mid, \smile, \left\{, \smile \right\} \right.$
$\quad\quad a \quad b \quad c \quad d \quad e$

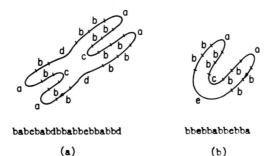

babcbabdbbabbcbbabbd   bbebbabbcbba

(a)   (b)

Fig. A.1. (a) Submedian chromosome; (b) telocentric chromosome.

231

and P:

⟨submedian chromosome⟩ → ⟨arm pair⟩ ⟨arm pair⟩
⟨telocentric chromosome⟩ → ⟨bottom⟩ ⟨arm pair⟩
⟨arm pair⟩ → ⟨side⟩⟨arm pair⟩
⟨arm pair⟩ → ⟨arm pair⟩ ⟨side⟩
⟨arm pair⟩ → ⟨arm⟩ ⟨right part⟩
⟨arm pair⟩ → ⟨left part⟩ ⟨arm⟩
⟨left part⟩ → ⟨arm⟩ c
⟨right part⟩ → c ⟨arm⟩
⟨bottom⟩ → b ⟨bottom⟩
⟨bottom⟩ → ⟨bottom⟩ b
⟨bottom⟩ → e
⟨side⟩ → b ⟨side⟩
⟨side⟩ → ⟨side⟩ b
⟨side⟩ → b
⟨side⟩ → d
⟨arm⟩ → b ⟨arm⟩
⟨arm⟩ → ⟨arm⟩ b
⟨arm⟩ → a

The arrowhead on the chromosome boundary in Fig. A.1 indicates the starting primitive and the direction of the string.

A syntax-directed translator, called the MOBILIZER [1, 2], is used to perform the syntax analysis of chromosome patterns. It recognizes the input pattern (represented by a string) as a chromosome or not a chromosome, and if a chromosome is recognized, as a particular kind of chromosome. Actually, the MOBILIZER works with numbers. The productions, which are coded correspondingly in terms of level numbers, are shown below.

$$P: \quad 77 \to 75, 75$$
$$76 \to 72, 75$$
$$75 \to 71, 75; \quad 75 \to 75, 71$$
$$75 \to 70, 73; \quad 75 \to 74, 70$$
$$74 \to 70, c$$
$$73 \to c, 70$$
$$72 \to b, 72; \quad 72 \to 72, b; \quad 72 \to e$$
$$71 \to b, 71; \quad 71 \to 71, b; \quad 71 \to d$$
$$70 \to b, 70; \quad 70 \to 70, b; \quad 70 \to a$$

The MOBILIZER, using a bottom-up procedure, starts with the first part of the pattern string and looks through the list of productions for a matching part. If, say, only the left component matches, then if the right component has a greater level number than the part of the object, an attempt is made to develop this part into the high generic form. The following example illustrates the procedure.

Step number	*babcbabdbabcbabd*
1	*b* 70 *bcbabdbabcbabd*
2	70 *bcbabdbabcbabd*
3	70 *cbabdbabcbabd*
4	74 *babdbabcbabd*
5	74 *b* 70 *bdbabcbabd*
6	74 70 *bdbabcbabd*
7	75 *bdbabcbabd*
8	75 *b* 71 *babcbabd*
9	75 71 *babcbabd*
10	75 *babcbabd*
11	75 *b* 70 *bcbabd*
12	75 70 *bcbabd*
13	75 70 *cbabd*
14	75 74 *babd*
15	75 74 *b* 70 *bd*
16	75 74 70 *bd*
17	75 75 *bd*
18	75 75 *b* 71
19	75 75 71
20	75 75
21	77

Using the same terminal set V_T as that used by Ledley, Lee and Fu [3] have constructed the following grammars for median, submedian, and acrocentric chromosomes (Fig. A.2), respectively.

(a) Grammar generating strings describing median chromosomes:

$$G_m = (V_{Nm}, V_{Tm}, P_m, S)$$

where $V_{Nm} = \{S, A, B, D, H, J, E, F\}$, $V_{Tm} = \{a, b, c, d\}$, and P_m:

$$
\begin{array}{lll}
S \to AA & D \to FDE & H \to a \\
A \to cB & D \to d & J \to a \\
B \to FBE & F \to b & \\
B \to HDJ & E \to b &
\end{array}
$$

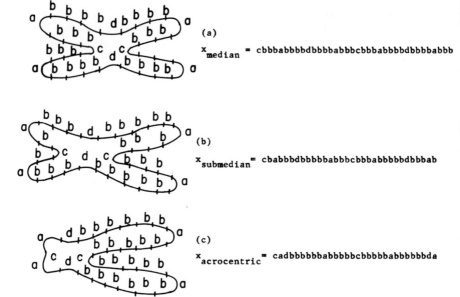

Fig. A.2. Three sample chromosomes: (a) median string representation; (b) submedian string representation; (c) acrocentric string representation.

(b) Grammar generating strings describing submedian chromosomes:

$$G_s = (V_{Ns}, V_{Ts}, P_s, S)$$

where $V_{Ns} = \{S, A, B, D, H, J, E, F, W, G, R, L, M, N\}$, $V_{Ts} = V_{Tm}$, and P_s:

$S \rightarrow AA$	$G \rightarrow FG$	$E \rightarrow b$
$A \rightarrow CM$	$L \rightarrow HNJ$	$F \rightarrow b$
$B \rightarrow FBE$	$L \rightarrow FL$	$G \rightarrow d$
$B \rightarrow FL$	$M \rightarrow FBE$	$H \rightarrow a$
$B \rightarrow RE$	$N \rightarrow FDE$	$J \rightarrow a$
$D \rightarrow FDE$	$R \rightarrow HNJ$	$W \rightarrow d$
$D \rightarrow FG$	$R \rightarrow RE$	
$D \rightarrow WE$	$W \rightarrow WE$	

(c) Grammar generating strings describing acrocentric chromosomes:

$$G_a = (V_{Na}, V_{Ta}, P_a, S)$$

where $V_{Na} = \{S, A, B, D, H, J, E, F, L, R, W, G\}$, $V_{Ta} = V_{Tm}$, and P_a:

$S \to AA$	$D \to WE$	$R \to RE$
$A \to cB$	$G \to FG$	$W \to WE$
$B \to FL$	$L \to HDJ$	$G \to d$
$B \to RE$	$F \to b$	$H \to a$
$E \to b$	$L \to FL$	$J \to a$
$D \to FG$	$R \to HDJ$	$W \to d$

References

1. R. S. Ledley, L. S. Rotolo, T. J. Golab, J. D. Jacobsen, M. D. Ginsburg, and J. B. Wilson, FIDAC: film input to digital automatic computer and associated syntax-directed pattern recognition programming system. In *Optical and Electro-Optical Information Processing* (J. T. Tippett *et al.*, eds.), Chapter 33, pp. 591–614. MIT Press, Cambridge, Massachusetts, 1965.
2. R. S. Ledley, *Programming and Utilizing Digital Computers*. McGraw-Hill, New York, 1962.
3. H. C. Lee and K. S. Fu, A stochastic syntax analysis procedure and its application to pattern classification. *IEEE Trans. Computers* **C-21**, 660–666 (1972).

Appendix B

PDL (Picture Description Language)

The PDL (Picture Description Language) developed by Shaw [1, 2] is briefly described in this appendix. Each primitive (terminal) is labeled at two distinguished points, a tail and a head. A primitive can be linked or concatenated to other primitives only at its tail and/or head. Four binary concatenation operators are defined.

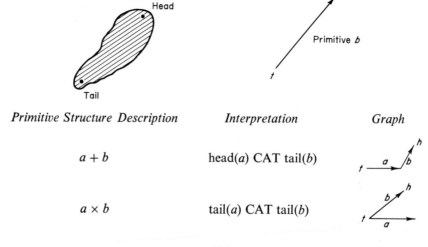

Primitive Structure Description	Interpretation	Graph
$a + b$	head(a) CAT tail(b)	
$a \times b$	tail(a) CAT tail(b)	

Primitive Structure Description	*Interpretation*	*Graph*

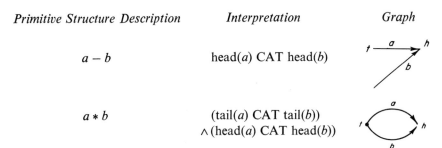

$a - b$	head(a) CAT head(b)	
$a * b$	(tail(a) CAT tail(b)) $\wedge$ (head(a) CAT head(b))	

A unary operator $\sim$ is defined as a tail/head reverser. For example,

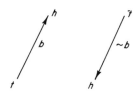

The grammar that generates sentences (PDL expressions) in PDL is a context-free grammar $G = (V_N, V_T, P, S)$, where $V_N = \{S, SL\}$, $V_T = \{b\} \cup \{+, \times, -, *, \sim, /, (,)\} \cup \{l\}$, b may be any primitive, (including the "null point primitive" λ, which has identical tail and head), and P:

$$S \to b, \quad S \to (S \varnothing_b S), \quad S \to (\sim S), \quad S \to SL, \quad S \to (/SL)$$
$$SL \to S^l, \quad SL \to (SL \varnothing_b SL), \quad SL \to (\sim SL), \quad SL \to (/SL)$$
$$\varnothing_b \to +, \quad \varnothing_b \to \times, \quad \varnothing_b \to -, \quad \varnothing_b \to *$$

l is a label designator which is used to allow cross reference to the expressions S within a description. The / operator is used to enable the tail and head of an expression to be arbitrarily located. The following examples (Figs. B.1–B.3) illustrate the operations of PDL for pattern structural description.

For the primitives defined in Fig. B.4, the PDL expressions for the upper-case English characters are listed in Fig. B.5.

In high-energy particle physics, one of the most common methods for obtaining the characteristics of an elementary particle is to analyze the trajectory "trail" left by the particle and its by-products in a detector chamber, such as a bubble or spark chamber. Several hundred thousand photographs of these trails or tracks might be taken in a typical experiment. Because of the large numbers involved and the accuracy and quantity of computation required for each interesting photograph, machine processing of the pictures is desirable.

Primitives:
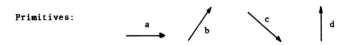

PDL expression describing character A:

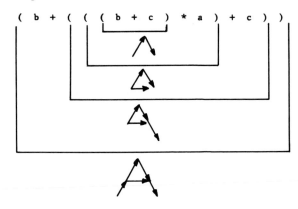

PDL expression describing character F:

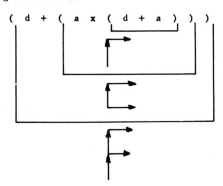

Fig. B.1. PDL structural description of the characters A and F.

(a) Pattern: A 4-node graph

Primitives:

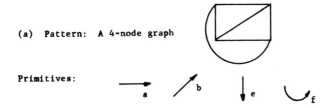

PDL expression describing the pattern:

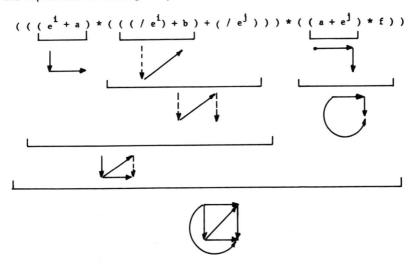

$$((((e^i + a) * (((/ e^i) + b) + (/ e^j))) * ((a + e^j) * f))$$

(b) Pattern: A 3-dimensional cube

Primitives:

a   b   d

PDL expression describing the pattern:

$$(((a * ((d^i + a) + (\sim d^j))) * ((((/ d^i) + b)$$
$$+ ((a * ((\sim d) + (a^k + d))) + (\sim b))) + (\sim (/ d^j))))$$
$$* ((b + (/ a^k)) + (\sim b)))$$

Fig. B.2. PDL descriptions of the patterns (a) a four-node graph and (b) a three-dimensional cube.

$G = (V_N, V_T, P, S)$

where

$V_N = \{S, A, House, Triangle\}$

$V_T = \{ \quad \underset{a}{\longrightarrow} \quad , \quad \nearrow^{b} \quad , \quad \searrow_{c} \quad , \quad \big\downarrow e \quad , \quad (\; , \;) \; , \; +, \; x, \; -, \; *, \; \sim \}$

and P:

S → A,   S → House

A → (b + (Triangle + c))

House → ((e + (a + (~ e))) * Triangle)

Triangle → ((b + c) * a)

L(G) = { (b+ (((b+c) *a) + c)) , ((e + (a + (~e))) * ((b+c) *a)) }

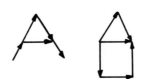

Fig. B.3. PDL structural description of the character A and the pattern "House."

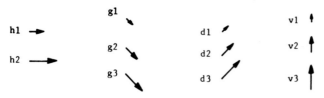

Fig. B.4. Pattern primitives.

$A \rightarrow (d2 + ((d2 + g2) * h2) + g2)$

$B \rightarrow ((v2 + ((v2 + h2 + g1 + (\sim(d1 + v1))) * h2) + g1 + (\sim(d1 + v1))) * h2)$

$C \rightarrow (((\sim g1) + v2 + d1 + h1 + g1 + (\sim v1)) \times (h1 + ((d1 + v1) \times \lambda)))$

$D \rightarrow (h2 * (v3 + h2 + g1 + (\sim(d1 + v2))))$

$E \rightarrow ((v2 + ((v2 + h2) \times h1)) \times h2)$

$F \rightarrow ((v2 + ((v2 + h2) \times h1)) \times \lambda)$

$G \rightarrow (((\sim g1) + v2 + d1 + h1 + g1 + (\sim v1)) \times (h1 + ((d1 + v1 - h1) \times \lambda)))$

$H \rightarrow (v2 + (v2 \times (h2 + (v2 \times (\sim v2)))))$

$I \rightarrow (v3 \times \lambda)$

$J \rightarrow ((((\sim g1) + v_1) \times h1) + ((d1 + v3) \times \lambda))$

$K \rightarrow (v2 + (v2 \times d2 \times g2))$

$L \rightarrow (v3 \times h2)$

$M \rightarrow (v3 + g3 + d3 + (\sim v3))$

$N \rightarrow (v3 + g3 + (v3 \times \lambda))$

$O \rightarrow (h1 * ((\sim g1) + v2 + d1 + h1 + g1 + (\sim(d1 + v2))))$

$P \rightarrow ((v2 + ((v2 + h2 + g1 + (d1 + v1))) * h2)) \times \lambda)$

$Q \rightarrow (h1 * ((\sim g1) + v2 + d1 + h1 + g1 + (\sim(d1 + ((\sim g1) \times g1) + v2))))$

$R \rightarrow (v2 + (h2 * (v2 + h2 + g1 + (\sim(d1 + v1))) + g2)$

$S \rightarrow ((((\sim g1) + v1) \times h1) + ((d1 + v1 + (\sim(g1 + h1 + g1)) + v1 + d1 + h1 + g1 + (\sim v1)) \times \lambda))$

$T \rightarrow ((v3 + (h1 \times (\sim h1))) \times \lambda)$

$U \rightarrow ((((\sim g1) + v3) \times h1) + ((d1 + v3) \times \lambda))$

$V \rightarrow ((\sim g3) \times d3 \times \lambda)$

$W \rightarrow (((\sim g3) + d3 + g3) + (d3 \times \lambda))$

$X \rightarrow (d2 + ((\sim g2) \times d2 \times g2))$

$Y \rightarrow ((v2 + ((\sim g2) \times d2)) \times \lambda)$

$Z \rightarrow ((d3 - h2) \times h2)$

Fig. B.5. PDL expressions of English characters.

In addition to the particle tracks, the pictures usually contain some identifying information (in a " data box "), such as frame number, view, input beam characteristics, and date, and a set of "fiducials," which are marks on the chamber whose positions are precisely known. Fiducials allow the tracks to be reconstructed in real space.

Figure B.6 gives the syntax for an abstracted particle physics picture. A negatively charged particle TM is assumed to enter a chamber containing positive particles P and under the influence of a magnetic field; TM enters from the left. The following types of reactions are provided for:

(a)   Interaction with P:

$$TM + P \rightarrow TM + TP$$
$$\rightarrow TM + TP + TN$$
$$\rightarrow TN$$

(b)   Negative particle decay

$$TM \rightarrow TM + TN$$

(c)   Neutral particle decay

$$TN \rightarrow TM + TP$$

(d)   Positive particle decay

$$TP \rightarrow TP + TN$$

TP and TN represent positively charged and neutral particles, respectively. The notation used above is similar to the conventional physics notation. The products of the reactions can themselves undergo the same series of reactions: this can occur an indefinite number of times. The chamber has four fiducials (X's) and an identification box.

The PDL descriptions are ordered for left-to-right recognition in that the lower left-hand fiducial FI appears first and its center is then used as the tail for the descriptions of the rest of the fiducials FID, the identification box ID, and the particle tracks PT. The starting primitive es is defined as a search strategy to find the lower arm of the left-corner fiducial. Blank (invisible) and "don't care" primitives connecting disjoint primitives are very useful for describing simple geometric relations, such as those occurring in the apparent stopping and the later appearance of particle tracks, and the " space " between adjacent characters and adjacent words. When a relation is to be described between disjoint primitives separated by other patterns, the separating patterns are defined as don't care primitives. The "null point" primitive λ is defined as a primitive with identical tail and head; hence it is represented as a

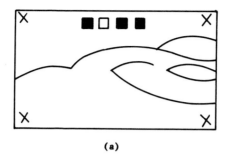

(a)

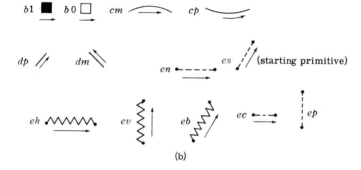

(b)

$\text{PICTURE} \rightarrow (es + (FI + (FID \times (ID \times PT))))$

$\quad FI \rightarrow (dp + (dm \times (dp \times (\lambda - dm))))$

$\quad FID \rightarrow ((eh + X) + ((ev + X)) - (X + eh)))$

$\quad ID \rightarrow ((eb + B) + ((ec + B) + ((ec + B) + (ec + B))))$

$\quad PT \rightarrow (ep + TM)$

$\quad X \rightarrow ((dp \times dm) \times ((\sim dp) \times (\lambda - dm)))$

$\quad B \rightarrow bo, B \rightarrow b1$

$\quad TM \rightarrow (cm + MD), TM \rightarrow (cm + MP), TM \rightarrow cm$

$\quad MP \rightarrow (P + ((TM \times TP) \times TN)), MP \rightarrow (P + (TM \times TP)), MP \rightarrow (P + TN)$

$\quad MD \rightarrow (TM \times TN), MD \rightarrow TM$

$\quad TP \rightarrow (cp + PD), TP \rightarrow cp$

$\quad TN \rightarrow (en + (N + (TM \times TP)))$

$\quad PD \rightarrow (TP \times TN), PD \rightarrow TP$

$\quad P \rightarrow \lambda$

$\quad N \rightarrow \lambda$

(c)

Fig. B.6. Syntax for a particle physics picture. (a) Sample picture; (b) primitives (– – – , blank; ⋀⋀ , "don't care"; λ, null point); (c) productions.

labeled node in a graph. By introducing λ, PDL can deal with points as well as edges.

For more application examples with more details, readers are referred to the work of Shaw [1, 2].

References

1. A. C. Shaw, *The Formal Description and Parsing of Pictures*. Rep. SLAC-84. Stanford Linear Accelerator Center, Stanford Univ., Stanford, California, 1968.
2. A. C. Shaw, A formal picture description scheme as a basic for picture processing systems. *Information and Control* **14**, 9–52 (1969).

Appendix C

Syntactic Recognition of Two-Dimensional Mathematical Expressions

The grammar used by Anderson [1] for the recognition of two-dimensional mathematics is described in the following:

$$G = (V_N, V_T, P, \langle \text{expression} \rangle)$$

where

$V_N = \{\langle \text{expression} \rangle, \langle \text{expression } 0 \rangle, \langle \text{expression } 1 \rangle, \langle \text{term} \rangle,$
$\langle \text{sumterm} \rangle, \langle \text{adjterm} \rangle, \langle \text{multerm} \rangle, \langle \text{factor} \rangle, \langle \text{intbody} \rangle,$
$\langle \text{limit} \rangle, \langle \text{variable} \rangle, \langle \text{subscriptlist} \rangle, \langle \text{subscriptlist } 0 \rangle,$
$\langle \text{trigname} \rangle, \langle \text{funcop} \rangle, \langle \text{trigop} \rangle, \langle \text{sumop} \rangle, \langle \text{addop} \rangle\}$

$V_T = \{\langle \text{letter} \rangle, \langle \text{unsigned: integer} \rangle, \langle \text{unsigned: number} \rangle,$

‖——, ——‖, { , }, ⊢——, ——⊣, [,], (,), ∶, ⁚, ,, =, ∞, ↓, ↑,

$\int, \sqrt{\ }, \prod, \sum, +, -, \cdot, /,$ sin, cos, tan, cot, sec,

csc, log, exp}

$\langle \text{letter} \rangle \in \{a, b, c, \ldots, z\}$
$\langle \text{unsigned: integer} \rangle \in \{\text{integers}\}$
$\langle \text{unsigned: number} \rangle \in \{\text{real numbers}\}$

245

and P:

(1) $\langle expression \rangle \rightarrow \langle expression\ 0 \rangle$
(2) $\langle expression\ 0 \rangle \rightarrow \langle expression\ 0 \rangle \langle addop \rangle \langle term \rangle$
(3) $\langle expression\ 0 \rangle \rightarrow \langle addop \rangle \langle term \rangle$
(4) $\langle expression\ 0 \rangle \rightarrow \langle expression\ 1 \rangle$
(5) $\langle expression\ 1 \rangle \rightarrow \langle term \rangle$
(6) $\langle term \rangle \rightarrow \langle adjterm \rangle \langle sumterm \rangle$
(7) $\langle term \rangle \rightarrow \langle sumterm \rangle$
(8) $\langle term \rangle \rightarrow \langle adjterm \rangle \langle term \rangle$
(9) $\langle term \rangle \rightarrow \langle adjterm \rangle$
(10) $\langle sumterm \rightarrow \int \vdash\!\!-\!\!- \langle limit \rangle; \langle limit \rangle -\!\!-\!\!\dashv \langle intbody \rangle$
(11) $\langle sumterm \rangle \rightarrow \int \vdash\!\!-\!\!- ; -\!\!-\!\!\dashv \langle intbody \rangle$
(12) $\langle sumterm \rangle \rightarrow \int \vdash\!\!-\!\!- \langle expression \rangle; \langle expression \rangle -\!\!-\!\!\dashv \langle intbody \rangle$
(13) $\langle sumterm \rangle \rightarrow \langle sumop \rangle \Vdash\!\!-\!\!- \langle variable \rangle = \langle expression \rangle;$
     $\langle expression \rangle -\!\!-\!\!\Vdash \langle term \rangle$
(14) $\langle sumterm \rangle \rightarrow \langle sumop \rangle \Vdash\!\!-\!\!- \langle variable \rangle: -\!\!-\!\!\Vdash \langle term \rangle$
(15) $\langle intbody \rangle \rightarrow \langle expression\ 1 \rangle\ d\ \langle variable \rangle$
(16) $\langle adjterm \rangle \rightarrow \langle adjterm \rangle \langle multerm \rangle$
(17) $\langle adjterm \rangle \rightarrow \langle multerm \rangle$
(18) $\langle multerm \rangle \rightarrow \langle trigop \rangle \langle multerm \rangle$
(19) $\langle multerm \rangle \rightarrow \langle funcop \rangle \langle multerm \rangle$
(20) $\langle multerm \rangle \rightarrow (\langle expression \rangle)/(\langle expression \rangle)$
(21) $\langle multerm \rangle \rightarrow \langle factor \rangle$
(22) $\langle factor \rangle \rightarrow \sqrt{}\ [\langle expression \rangle; \langle expression \rangle]$
(23) $\langle factor \rangle \rightarrow \sqrt{}\ [; \langle expression \rangle]$
(24) $\langle factor \rangle \rightarrow (\langle expression \rangle)$
(25) $\langle factor \rangle \rightarrow \langle variable \rangle$
(26) $\langle factor \rangle \rightarrow \langle unsigned: integer \rangle$
(27) $\langle factor \rangle \rightarrow \langle unsigned: number \rangle$
(28) $\langle factor \rangle \rightarrow \langle factor \rangle \uparrow [\langle expression \rangle]$
(29) $\langle funcop \rangle \rightarrow \log$
(30) $\langle funcop \rangle \rightarrow \exp$
(31) $\langle limit \rangle \rightarrow \langle addop \rangle \infty$
(32) $\langle limit \rangle \rightarrow \infty$
(33) $\langle variable \rangle \rightarrow \langle letter \rangle \downarrow \{\langle subscriptlist \rangle\}$
(34) $\langle variable \rangle \rightarrow \langle letter \rangle$
(35) $\langle subscriptlist \rangle \rightarrow \langle subscriptlist\ 0 \rangle$
(36) $\langle subscriptlist\ 0 \rangle \rightarrow \langle subscriptlist\ 0 \rangle, \langle expression \rangle$
(37) $\langle subscriptlist\ 0 \rangle \rightarrow \langle expression \rangle$
(38) $\langle trigop \rangle \rightarrow \langle trigname \rangle \uparrow [\langle expression \rangle]$
(39) $\langle trigop \rangle \rightarrow \langle trigname \rangle$

(40) $\langle \text{trigname} \rangle \rightarrow \sin$
(41) $\langle \text{trigname} \rangle \rightarrow \cos$
(42) $\langle \text{trigname} \rangle \rightarrow \tan$
(43) $\langle \text{trigname} \rangle \rightarrow \cot$
(44) $\langle \text{trigname} \rangle \rightarrow \sec$
(45) $\langle \text{trigname} \rangle \rightarrow \csc$
(46) $\langle \text{addop} \rangle \rightarrow +$
(47) $\langle \text{addop} \rangle \rightarrow -$
(48) $\langle \text{sumop} \rangle \rightarrow \sum$
(49) $\langle \text{sumop} \rangle \rightarrow \prod$

This syntax has many interesting features. Some of the more unusual productions rules are described next.

Rule	Discussion
6	Handles cases like $3 \sum_{i=1}^{n} x_i$. Note that syntactic units like $\sum_{i=1}^{n} x_i$ are "syntactically unsymmetric"; to the left they act like a factor, capable of implied multiplication; to the right, they act like a term, being delimited only by an addition operator.
14	Handles the configuration $\sum_i x_i$, where the limits of the summation are assumed to extend over the length of x.
18	This rule, with rule 16, handles the peculiarities of trigonometric notation. It allows the configuration "$\sin x \cos y$" to be parsed as $\sin (x) * \cos (y)$.

The grammar is a simple precedence grammar [2], and its precedence matrix is shown in Fig. C.1. The explicit characters $A, C, D, E, G, I, L, N, O, P, S, T, X$ in Fig. C.1 are used in trigonometric and function names. There are three nonunique precedence relations in the matrix; namely,

$$C \gtreqqless S, \qquad S \gtreqqless C, \qquad S \gtreqqless E$$

This means that it is necessary to look farther than one symbol to the right to determine, for example, whether the substring "cs" occurs as part of the string "sec sin," or whether it occurs as part of the string "csc."

The recognition algorithm is shown in Fig. C.2. Assume that hand-printed characters can be recognized† and, in addition, that each character's position is described by the six spatial coordinates $x_{\min}, x_{\text{center}}, x_{\max}, y_{\min}, y_{\text{center}},$ and

† For on-line recognition of hand-printed characters, see, for example, the literature [3–5].

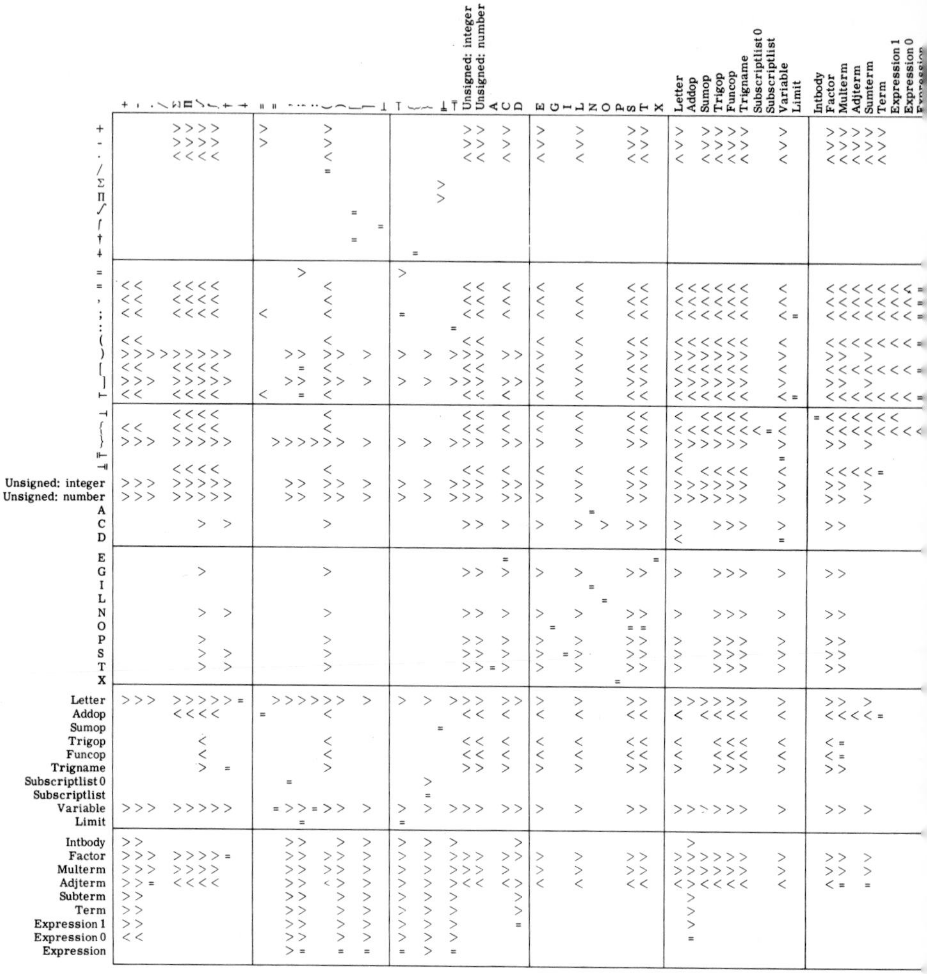

Fig. C.1. Precedence matrix for mathematical syntax.

y_{max}. (See Fig. C.3.) The following conditions are placed on "well-formed" expressions:

(1) The expressions above and below a horizontal line (indicating division) are bounded in the x direction by the extent of the line.

(2) The upper and lower limits accompanying a summation or product sign are bounded in the x direction by the extent of the sign.

(3) Two adjacent characters, c_1 and c_2, which are meant to be on the same typographical line, must satisfy the conditions:

$$|y_{center}(c_1) - y_{center}(c_2)| < h_{tolerance}$$

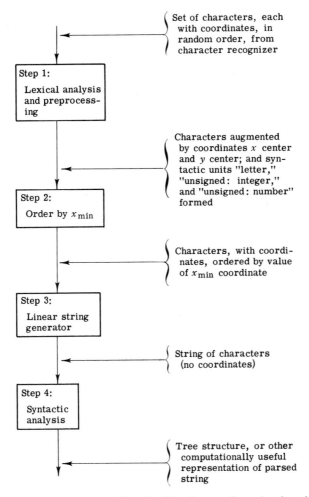

Fig. C.2. The four-step recognition algorithm for two-dimensional mathematics.

and

$$(x_{\min}(c_2) - x_{\max}(c_1)) < h_{\max}$$

where $h_{\text{tolerance}}$ and $h_{\max}$ are parameters that may be adjusted by the system user. That is, their typographical centers must be sufficiently similar and the space between them must not be too great.

(4)   Each character c in a limit on an integral sign must be placed so that

$$|y_{\text{center}}(c) - y_{\text{center}}(\smallint)| > h_{\text{tolerance}}$$

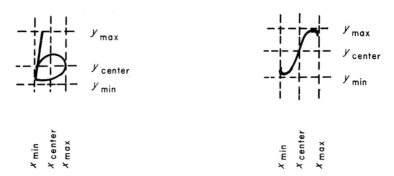

Fig. C.3. Spatial coordinates of characters.

that is, it must not be on the same typographical line as the integral sign. Characters above $y_{center}(\int)$ will be assumed to be part of the upper limit; characters below will be part of the lower limit. The limits must also not extend farther to the left than $x_{min}(\int)$; all characters in the body of the integral must be to the right of all characters in the limits.

(5) The characters giving the value of a root must be bounded in the x direction by the extent of the character $\sqrt{\,}$, and the expression within the root sign must be bounded in the x direction by the horizontal line of the root, and in the y direction by the extent of the $\sqrt{\,}$.

(6) All characters in a subscript or superscript must be sufficiently below or above (respectively) the center of the letter or factor to which they apply so that they are not on the same typographical line as that letter or factor. All characters in a subscript or superscript on a character c must be to the left of the next character on the same typographical line as c.

These restrictions can be encoded in the syntax rules. They still allow limits on an integration to be placed above and below, or to the right of, the integral sign; and they still allow a "tuning" of the recognition by the adjustment of parameters.

With the foregoing restrictions, mathematical notation has an important property: the leftmost character in any major syntactic unit (expression, term, factor, etc.) defines the typographical center in the y direction of that syntactic unit. The utility of this property may be seen from the following example. Consider the character configuration

$$\int_{i-1}^{i+1} \frac{\sqrt{x^2 - 1}}{\sqrt{x^2 + 1}}\, dx$$

If we order the characters by their x_{min} coordinate (in increasing order), then we know that the *first* character in the ordered set (in this case, the integral sign) defines the typographical center of the whole configuration. Moreover, the next character in the set whose y_{center} is sufficiently close to y_{center} of the integral sign to be on the same typographical line (in this case, the horizontal line) is the first character in the body of the integral; and if it is an operator, it is the primary operator in that body (since the primary operator defines the center of the syntactic unit).

As shown in Fig. C.2, the first step, lexical analysis, was discussed as "preprocessing" earlier. The second step is simply a sort. It should be noted that we are assuming that the mathematical expression was written on one "line." If this is not the case, it will be necessary for the preprocessor in step 1 to look for gross changes in the y coordinates in the character stream and to normalize them so that the characters appear to be on the same line. (This would not be difficult, since it again may be assumed that the first character written on the new line is the leftmost one, and that it defines the typographical center of that line; the difference between the new y_{center} and that of the previous line should be added to each y coordinate of the new character and to the succeeding ones for normalization, and their x coordinates should be incremented appropriately to create a virtual "line" of indefinite length to the right of the first line.)

Step 3, the linear string generator, performs tests on the coordinates of the characters and emits special characters which specify relationships among the intput characters. The relationships for which this program tests are the following five deviations from linearity: (1) division, indicated by a horizontal line with expressions above and below the line; (2) limits on summation $\sum$, product $\prod$, and integration $\int$; (3) the use of an integer or variable to show the value of a root, for example, $\sqrt[5]{a + e}$; (4) subscripts; (5) superscripts. For example, step 3 changes the input configuration

$$a^2$$

into the string $a \uparrow [2]$; the configuration

$$\frac{a + b}{e}$$

becomes the string $(a + b)/(e)$. The string of characters (with coordinates no longer necessary), converted from a set of characters (with positional coordinates) ordered by their x_{min} values, can then be parsed by the simple precedence grammar.

An extension of precedence analysis to the two-dimensional case has been carried out recently by Chang [6]. For one-dimensional string languages, precedence analysis is a technique for ordering the operators of a linear string so that a structure can be constructed for that string (Chapter 4). Similarly, the two-dimensional extension is a technique for ordering the operators of a two-dimensional pattern so that a structure can be constructed for that pattern. In addition to operator precedence, the concept of operator dominance is introduced. One operator dominates another if and only if the latter is in the range of the former and the converse is false. Operator dominance is a way to take into account the geometric relationship between operators. Combining the concepts of operator precedence and operator dominance, we can define an ordering relation on the operators of a two-dimensional pattern. A structure (syntax) can then be constructed for that pattern.

References

1. R. H. Anderson, *Syntax-Directed Recognition of Hand-Printed Two-Dimensional Mathematics*. Ph.D. Thesis, Div. of Eng. and Appl. Phys., Harvard Univ., Cambridge, Massachusetts, 1968; also appears in summary form in *Interactive Systems for Experimental Applied Mathematics*, (M. Klerer and J. Reinfelds, eds.), pp. 436–459. Academic Press, New York, 1968.
2. N. Wirth and H. Weber, EULER: A generalization of ALGOL and its formal definition, Pt. I. *Comm. ACM* **9**, 13–25 (1966).
3. G. F. Groner, Real-time recognition of hand-printed text. *Proc. AFIPS Fall Joint Comput. Conf., 1966*, p. 591.
4. M. I. Bernstein and H. L. Howell, *Hand-Printed Input for On-Line Systems*. Rep. TM-3937. Systems Develop. Corp., Santa Monica, California, 1968.
5. G. H. Miller, On-line recognition of hand-generated symbols. *Proc. AFIPS Fall Joint Comput. Conf., 1969*, p. 399.
6. S. K. Chang, A method for the structural analysis of two-dimensional mathematical expressions. *Information Sci.* **2**, 253–272 (1970).
7. W. A. Martin, Computer input/output of mathematical expressions. *Proc. Symp. Symbolic and Algebraic Manipulation, 2nd, Los Angeles, California, March 23–25, 1971*, pp. 78–89.

Appendix D

Syntactic Description of Hand-Printed FORTRAN Characters

The grammar proposed by Narasimhan [1], which generates a language to describe FORTRAN characters, is defined as

$$G = (V_T, \ Y, R, P, T)$$

where V_T is a set of primitives, Y a set of attributes, R a set of relations, P a set of composition rules, and T is a set of transformations.

V_T and Y together allow us to generate primitives with assigned properties (i.e., with specified attribute values). Let a_1, a_2 be primitives with assigned properties, and let $r \in R$ be a binary relation defined over some subset of the attribute values of a_1 and a_2. $r(a_1, a_2)$ defines a subpattern A whose constituents are the primitives a_1 and a_2, and whose assigned properties satisfy the relationship r. Then

$$A \to r(a_1, a_2)$$

is a composition rule. In general, the composition rule assumes the form

$$A_3 \to r(A_1, A_2), \qquad r \in R$$

where A_1, A_2 may be either subpatterns or primitives.

The basic frame with primitive regions is shown in Fig. D.1. A region is a set of connected points and can be considered as a convex neighborhood

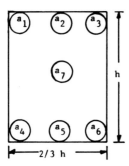

Fig. D.1. Basic frame with primitive regions.

(e.g., a circular neighborhood). The size of a region, that is, the number of points in it, is at least 1, but usually much larger. In Fig. D.1, seven such regions are identified by the names a_1, a_2, a_3, a_4, a_5, a_6, a_7. The following is a list of relations with brief explanations or illustrations:

(1) $VT(a_i)$ Vertical through a_i

(2) $HZ(a_i)$ Horizontal through a_i

(3) $RT(a_i)$, $i \neq 7$ Join a_i by straight line to rightmost point on opposite edge.

(4) $LT(a_i)$, $i \neq 7$ Same as (3) with left replacing right.

(5) $ST(a_i, a_j)$ Join a_i, a_j by straight line.

(6) $CE(a_i, a_j)$, C curve

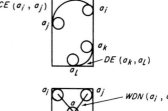

(7) $DE(a_i, a_j)$, D curve

(8) $W\alpha(a_i, a_j; a)$, $\alpha = LT$ for left, RT for right, UP for above, DN for below

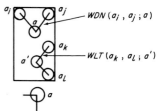

(9) $HK(a)$, hook centered on a

(10) $BB(a)$ A degenerated line and it identifies a connected set of points around a; the size is roughly from 1 to 9.

Each line segment has certain distinguished regions and a thickness or width parameter on it as attributes. In the case of lines specified by only one region, the length parameter is an attribute. If this is not explicitly specified, the line segment is assumed to extend from edge to edge of the frame. The complete set of composition rules for FORTRAN characters is given in Fig. D.2. The following notations are used:

$+$ concatenation or sequence control

$\bar{a}$ slightly above a

$\underline{a}$ slightly below a

$<$ less than full size (refer to segment length)

' '   character name identifier.

$$\text{'A'} \rightarrow LT(a_2; a) + RT(a_2; b) + ST(a, b)$$
$$\text{'B'} \rightarrow VT(a_1; a) + DE(a_1; a) + DE(a, a_4)$$
$$\text{'C'} \rightarrow CE(\underline{a}_3, \bar{a}_6)$$
$$\text{'D'} \rightarrow VT(a_1) + DE(a_1, a_4)$$
$$\text{'E'} \rightarrow \text{'F'} + HZ(a_4)$$
$$\text{'F'} \rightarrow VT(a_1; a) + HZ(a_1) + HZ(a; <)$$
$$\text{'G'} \rightarrow \text{'C'} + HK(\bar{a}_6)$$
$$\text{'H'} \rightarrow VT(a_1; a) + VT(a_3; b) + ST(a, b)$$
$$\text{'I'} \rightarrow VT(a_2) + HZ(a_2; <) + HZ(a_5; <)$$
$$\text{'J'} \rightarrow ST(a_2; a_7) + DE(a_7; \bar{a}_4) + HZ(a_2; <)$$
$$\text{'K'} \rightarrow VT(a_1; a) + WLT(a_3, a_6; a)$$
$$\text{'L'} \rightarrow VT(a_1) + HZ(a_4)$$
$$\text{'M'} \rightarrow VT(a_1) + WDN(a_1, a_3) + VT(a_3)$$
$$\text{'N'} \rightarrow VT(a_1) + RT(a_1) + VT(a_6)$$
$$\text{'Ø'} \rightarrow \text{'0'} + \text{'1'}$$
$$\text{'P'} \rightarrow VT(a_1; a) + DE(a_1, a)$$
$$\text{'Q'} \rightarrow \text{'0'} + ST(a_7, a_6)$$
$$\text{'R'} \rightarrow \text{'P'} + ST(a, a_6)$$
$$\text{'S'} \rightarrow CE(a_3, a_7) + DE(a_7, \bar{a}_4)$$
$$\text{'T'} \rightarrow VT(a_2) + HZ(a_2)$$
$$\text{'U'} \rightarrow ST(a_1, \bar{a}_4) + CE(\bar{a}_4, \bar{a}_6) + ST(\bar{a}_6, a_3)$$
$$\text{'V'} \rightarrow ST(a_1, a_5) + ST(a_5, a_3)$$
$$\text{'W'} \rightarrow VT(a_1) + WUP(a_4, a_6) + VT(a_6)$$
$$\text{'X'} \rightarrow LT(a_3) + RT(a_1)$$
$$\text{'Y'} \rightarrow ST(a_1, a_7) + LT(a_3)$$
$$\text{'Z'} \rightarrow HZ(a_1) + LT(a_3) + HZ(a_4)$$
$$\text{'0'} \rightarrow CE(a_2, a_5) + CE(a_5, a_2)$$
$$\text{'1'} \rightarrow VT(a_2)$$

Fig. D.2. Composition rules for FORTRAN characters. Continued on following page.

'2' $\rightarrow DE(g_1, a_7) + CE(a_7, a_4) + HZ(a_4)$

'3' $\rightarrow HZ(a_1) + ST(a_3, a_7) + DE(a_7, a_4)$

'4' $\rightarrow ST(a_1, \bar{a}_4) + HZ(\bar{a}_4) + ST(a_7, a_5)$

'5' $\rightarrow HZ(a_1) + ST(a_1, \bar{a}_4) + DE(\bar{a}_4, a_7) + DE(a_7, a_4)$

'6' $\rightarrow CE(g_3, \bar{a}_4) + CE(\bar{a}_4, a_7) + CE(a_7, \bar{a}_4)$

'7' $\rightarrow HZ(a_1) + ST(a_3, a_5)$

'8' $\rightarrow CE(a_2, a_7) + DE(a_7, a_5) + DE(a_5, a_7) + CE(a_7, a_2)$

'9' $\rightarrow DE(g_3, a_7) + DE(a_7, g_3) + DE(g_3, \bar{a}_4)$

'+' $\rightarrow HZ(a_7) + VT(a_7)$

'−' $\rightarrow HZ(a_7)$

'*' $\rightarrow HZ(a_7) + VT(a_7) + RT(a_1) + LT(a_3)$

'=' $\rightarrow HZ(\bar{a}_7) + HZ(g_7)$

'/' $\rightarrow LT(a_3)$

'≠' $\rightarrow$ '=' + '/'

'·' $\rightarrow BB(a_5)$

',' $\rightarrow BB(\bar{a}_5) + DE(\bar{a}_5, a_4)$

'(' $\rightarrow CE(a_2, a_5)$

')' $\rightarrow DE(a_2, a_5)$

'\$' $\rightarrow$ 'S' + $VT(a_7)$

Fig. D.2 Continued from preceding page.

If the original characters and their mirror images about a vertical axis through a_7 are considered, a transformation called MIRROR INVERT can be defined, using the following set of replacement rules.

$$a_1 \longleftrightarrow a_3$$
$$a_4 \longleftrightarrow a_6$$
$$LT \longleftrightarrow RT$$
$$CE \longleftrightarrow DE$$

Reference

1. R. Narasimhan, On the description, generation, and recognition of classes of pictures. In *Automatic Interpretation and Classification of Images* (A. Grasselli, ed.). Academic Press, New York, 1969.

Appendix E

Syntactic Recognition of Chinese Characters

A number of authors have proposed the use of a syntactic approach in the generation and/or recognition of Chinese characters [1–5]. In this appendix, the method used by Stallings [1] for recognition of printed Chinese characters is briefly described.

Pattern Primitives. Stroke segments.

Subpatterns. Components of a Chinese character described in terms of stroke segments.

Segmentation of a Chinese Character and Corresponding Relations between Subpatterns.

(i) East–West

Relation "left of"

(ii) North–South

Relation "above"

(iii) Border–Interior

Relation "surround"

(iv) Embedding of (i), (ii), or (iii) in one of the subregions of (i), (ii), or (iii). For example,

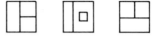

Structure. The structure of a Chinese character in terms of subpatterns and binary relations between subpatterns is represented by a tree. For example, the structure representation of the character 沁 is shown in Fig. E.1.

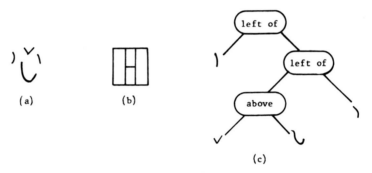

Fig. E.1. (a) The character; (b) segmentation; (c) tree representation.

Subpattern Analysis. The following computer programs are used for the analysis of character components.

(1) CONTOUR TRACING—to determine the boundary of each character component.

(2) SEARCH—to find some stroke segment to be used as a start point. (It is unimportant which particular segment of a component is found.)

(3) CRAWL—to "crawl along" a stroke segment in a given direction, halting when a node is encountered, that is, when an intersection or a tip of a stroke is reached. (The octal code is used to represent the direction of a stroke segment.)

(4) BUILD—to construct a graph to represent the component in terms of stroke segments.

The organization of the programs is shown in Fig. E.2, and an illustrative example is given in Fig. E.3.

Fig. E.2. Component analysis programs.

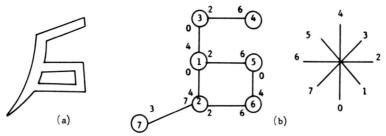

Fig. E.3. (a) The component and (b) graph representation and octal code. (The label on a branch at a node indicates the direction of that branch in terms of the octal code.)

Subpattern Recognition. A numeric code is generated for each component in a character. The code for a component is generated from its graph representation. The generating algorithm obeys the following rules.

(1) Start at the node in the upper left-hand corner of the graph. Exit by the branch with the lowest-valued label. Mark the exiting branch to indicate its having been taken, and write down the branch label.

(2) Upon entering a node, check to see if it is being visited for the first time. If so, mark the entering node to indicate this.

(3) Upon leaving a node, if there are available unused directions other than along the first entering branch, choose the one among these with the lowest-valued label. Leave by the first entering branch only as a last resort. Mark the exiting branch to indicate its having been taken and write down the label on the branch.

By this algorithm, all branches are traversed exactly once in each direction, so all labels are picked up. The code consists of the branch labels in the graph written down in the order in which they are encountered. An illustrative example is given in Fig. E.4. Recognition of a subpattern is accomplished

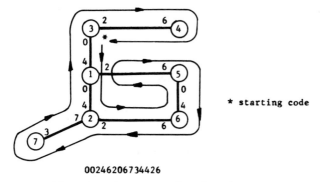

* starting code

00246206734426

Fig. E.4. Code representation of a subpattern.

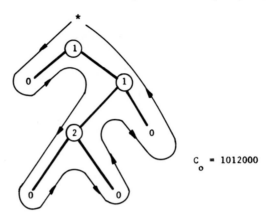

$C_o = 1012000$

Fig. E.5. Code representation of a tree.

by matching the code representing the unknown subpattern with the codes specifying various possible subpatterns.

Recognition of Characters. Each character is represented by a sequence of $n + 1$ codes where n is the number of components in the character.

$$C_0, C_1, \ldots, C_n$$

C_0 is the code generated from the tree representation of the character (Fig. E.1c), and C_1 through C_n are the codes of the components listed according to the order in which the components appear in code C_0.

C_0 is generated by traveling around the tree counterclockwise, starting from the root node of the tree, and picking up nodes and terminals (primitives or subpatterns) the first time they are encountered. The code used is 0 for terminals; 1 for " left of " node; 2 for " above " node; 3 for " surround " node. Figure E.5 shows the generation of code from the tree of Fig. E.1c. Recognition of a character is based on the matching between the sequence of $n + 1$ codes representing the unknown input and the codes specifying different classes of Chinese characters.

References

1. W. 'W. Stallings, Recognition of printed Chinese characters by automatic pattern analysis." *Comput. Graphics and Image Process.* **1**, 47–65 (1972).
2. O. Fujimura and Y. Kagaya, *Structure Patterns of Chinese Characters.* Annu. Bull. No. 3, pp. 131–148. Res. Inst. of Logopedics and Phoniatrics, Univ. of Tokyo, Japan, April 1968–July 1969.
3. T. Sakai, M. Nagao, and H. Terai, A description of Chinese characters using subpatterns. *Information Processing in Japan* **10** (1969).

4. B. K. Rankin and D. Tan, Component combination and frame-embedding in Chinese character grammars. NBS *Tech. Note* **492** (1970).
5. S. K. Chang, An interactive system for Chinese character generation and text editing. *Proc. IEEE Int. Conf. Cybernet. and Soc., Washington, D.C., October 9–12, 1973.*

Appendix F

Syntactic Recognition of Spoken Words

The application of syntactic approach to the recognition of sounds, words, and continuous speech has recently received increasing attention [1–4]. The method used by De Mori [1] for recognition of spoken digits is summarized in this appendix. Zero crossings are used to characterize each segment of the incoming speech signal [5, 6]. A careful analysis of the sequences of zero-crossing intervals from many spoken words indicated that intervals of 20-msec length provide meaningful short time statistics of the intervals. Inspection of these statistics leads to the conclusion that the zero-crossing intervals in each 20-msec segment can be classified into a few groups. The number of the zero-crossing intervals classified into the groups during a segmentation interval can then be used as features of the speech segment. Consequently, the segmentation of each pattern (speech signal) is based on 20-msec time intervals and the range of the zero-crossing intervals of the outputs of the low-pass filter (LPF) and high-pass filter (HPF), respectively, is subdivided according to Table F.1. A simple block diagram of the pre-processing part of the recognition system is shown in Fig. F.1.

For the nth segment of 20-msec length, compute

$$R_L(nT) = \{R_{L1}(nT), R_{L2}(nT), \ldots, R_{L7}(nT\}$$
$$R_H(nT) = \{R_{H1}(nT), \ldots, R_{H4}(nT)\}, \qquad T = 20 \text{ msec}$$

where $R_{ki}(nT)$ is the number of zero-crossing intervals from the kth filter

TABLE F.1

SPECIFICATIONS FOR THE GROUPS OF
ZERO-CROSSING INTERVALS[a]

	7 groups for LPF[b]	4 groups for HPF[b]
t_0	7.0	0.9
t_1	3.0	0.6
t_2	1.6	0.4
t_3	1.2	0.3
t_4	1.0	0.1
t_5	0.8	
t_6	0.6	
t_7	0.4	

[a] An incoming zero-crossing interval of duration t from the output of the low-pass filter (or the high-pass filter) is assigned to the group (L, i) (or (H, i)).
[b] Units msec.

$(k = L$ for LPF, $k = H$ for HPF) assigned to the group (L, i) or (H, i) during the nth segmentation period. From this information, then calculate the following parameters:

$$B_L(nT) = \frac{\sum_{i=1}^{7} (i - 1)R_{Li}(nT)}{\sum_{i=1}^{7} R_{Li}(nT)} \quad \text{(quantized into 60 levels)}$$

$$B_H(nT) = \frac{\sum_{i=1}^{4} (i - 1)R_{Hi}(nT)}{\sum_{i=1}^{4} R_{Hi}(nT)} \quad \text{(quantized into 30 levels)}$$

Thus, each spoken word can be represented pictorially on the B_L–B_H plane. Figure F.2 shows such a plot for the Italian spoken digit "nine" (NOVE).

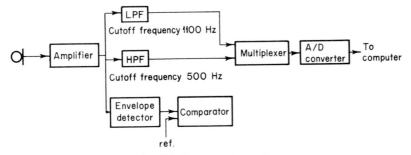

Fig. F.1. Preprocessing system.

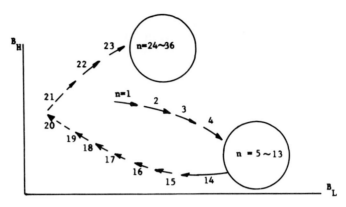

Fig. F.2. B_L–B_H representation of NOVE.

A relatively dense set of points is generated by the pronounciation of a vowel or a semivowel.

Syntactic Representation of a Spoken Word (in B_L–B_H Plane)

Primitives (Terminals)

(1) Silence interval (SL)—a sequence of points in the B_L–B_H plane whose coordinates are all zeros following at least one point whose coordinates are both nonzero. These points are labeled as

$$N \cdot p_1 \cdot p_2$$

where N is the symbol used for (SL), and p_1 and p_2 denote the starting position and the duration of the (SL), respectively. For example, $N \cdot 1010 \cdot 11$ represents an (SL) starting from the tenth segmentation interval after the beginning of the word for a duration of three successive segments.

(2) Stable zone (SZ)—a set of points representing successive segments of the word lies within a relatively small region and the number of points is higher than specified threshold. The (SZ) is described by

$$S \cdot s_1 \cdot s_2 \cdot s_3 \cdot s_4$$

where s_1 and s_2 have the same meaning as p_1 and p_2 for (SL), s_3 and s_4 are the coordinates of the center of gravity of the (SZ).

(3) Lines (LN)—nonstationary portions of the acoustics waveform approximated by straight-line segments. The (LN) is described by

$$L \cdot l_1 \cdot l_2 \cdot l_3 \cdot l_4$$

where l_1 and l_2 have the same meaning as p_1 and p_2 for (SL), l_3 is the length of the line segment, and l_4 is the slope of the line represented in octal code.

Nonterminals

(1) Silence fragment (NF)—an (SL) with the following description:

$$N \cdot n_1 \cdot n_2 \cdot 0 \cdot 0 \cdot 0 \cdot 0 \cdot 0$$

where n_1 and n_2 have the same meaning as p_1 and p_2. In terms of a production rule, (NF) is defined as

$$(NF) \rightarrow (SL)$$

(2) O fragment (OF)—a stable zone which cannot be composed by other primitives. Its description is

$$O \cdot o_1 \cdot o_2 \cdot o \cdot o \cdot o \cdot o_6 \cdot o_7$$

where o_1, o_2, o_6, and o_7 have the same meaning as s_1, s_2, s_3, and s_4, respectively. The production is

$$(OF) \rightarrow (SZ)$$

For example, a vowel between two silences usually leads to an (OF).

(3) I fragment (IF)—a picture ending with an (SZ) preceded by an (LN), an (SZ), or an (SZ) followed by an (LN). In production form,

$$(IF) \rightarrow \gamma(SZ)$$

where $\gamma \rightarrow (LN)$, $\gamma \rightarrow (SZ)$, or $\gamma \rightarrow (SZ)(LN)$. The (IF) description is

$$I \cdot i_1 \cdot i_2 \cdot i_3 \cdot i_4 \cdot i_5 \cdot i_6 \cdot i_7$$

where i_1 and i_2 have the same meaning as o_1 and o_2, and i_6 and i_7 are the coordinates of the center of gravity of the last (SZ). i_3 is a composition code defined by the following table:

(IF) composition	i_3 Code
(LN)(SZ)	01
(SZ)(SZ)	11
(SZ)(LN)(SZ)	101

If $i_3 = 11$, i_5 represents the slope (in octal code) of the line segment joining the two (SZ)'s, and i_4 denotes the length of the line segment joining the centers of gravity of the two (SZ)'s divided by 2^3.

(4) V fragment (VF)—a picture composed by two lines and terminated by an (SZ); that is,

$$(VF) \rightarrow \gamma^2(SZ)$$

A (VF) description is

$$V \cdot v_1 \cdot v_2 \cdot v_3 \cdot v_4 \cdot v_5 \cdot v_6 \cdot v_7$$

where v_1, v_2, v_6, and v_7 are similar to i_1, i_2, i_6, and i_7 of the (IF), v_3 is the composition code with a value 1 for (SZ) and 0 for (LN). v_5 is the sequence of slopes of the line segments each one of which is expressed in octal code. v_4 is the sequence of lengths of the line segments divided by 2^3.

(5) Z fragment (ZF)—a picture with three lines.

$$(ZF) \rightarrow \gamma^3(SZ)$$

A (ZF) description is

$$Z \cdot z_1 \cdot z_2 \cdot z_3 \cdot z_4 \cdot z_5 \cdot z_6 \cdot z_7$$

where $z_1, \ldots, z_7$ have the same meaning as $v_1, \ldots, v_7$ of a (VF).

Grammar: $G = (V_N, V_T, P, S)$, where

$V_N = \{$(NF), (OF), (IF), (VF), (ZF), γ, S, ZERO, ONE, TWO, THREE, FOUR, FIVE, SIX, SEVEN, EIGHT, NINE$\}$

$V_T = \{$(SL), (LN), (SZ)$\}$

P: $S \rightarrow ZERO,$ $S \rightarrow ONE,$ $S \rightarrow TWO,$ $S \rightarrow THREE$
          $S \rightarrow FOUR,$ $S \rightarrow FIVE,$ $S \rightarrow SIX,$ $S \rightarrow SEVEN$
          $S \rightarrow EIGHT,$ $S \rightarrow NINE,$
     (NF) $\rightarrow$ (SL),   (OF) $\rightarrow$ (SZ),   (IF) $\rightarrow \gamma$(SZ),
     (VF) $\rightarrow \gamma^2$(SZ),   (ZF) $\rightarrow \gamma^3$(SZ),
          $\gamma \rightarrow$ (LN), $\gamma \rightarrow$ (SZ), $\gamma \rightarrow$ (SZ)(LN)

ZERO $\rightarrow$ (IF) $\{0, \leq i_4 \leq 1, 6 \leq i_5 \leq 7, 32 \leq i_6 \leq 48, 0 \leq i_7 \leq 16\}$

ZERO $\rightarrow$ (VF) $\{00 \leq v_4 \leq 22, 26 \leq v_5 \leq 47, 32 \leq v_6 \leq 48,$
                        $0 \leq v_7 \leq 16\}$

ZERO $\rightarrow$ (ZF) $\{000 \leq z_4 \leq 111, 026 \leq z_5 \leq 047, 32 \leq z_6 \leq 48,$
                        $0 \leq z_7 \leq 16\}$

ONE $\rightarrow$ (VF) $\{00 \leq v_4 \leq 22, 06 \leq v_5 \leq 27, 32 \leq v_6 \leq 48,$
                        $0 \leq v_7 \leq 16\}$

ONE $\rightarrow$ (ZF) $\{000 \leq z_4 \leq 222, 072 \leq z_5 \leq 274, 32 \leq z_6 \leq 48,$
                        $0 \leq z_7 \leq 16\}$

TWO $\rightarrow$ (IF) $\{1 \leq i_4 \leq 2, 1 \leq i_5 \leq 2, 32 \leq i_6 \leq 40,$
                        $16 \leq i_7 \leq 24\}$

TWO $\rightarrow$ (VF) $\{00 \leq v_4 \leq 22, 01 \leq v_5 \leq 13, 32 \leq v_6 \leq 40,$
                        $16 \leq v_7 \leq 24\}$

THREE $\rightarrow$ (OF) $\{24 \leq o_6 \leq 40, 16 \leq o_7 \leq 24\}$

THREE $\rightarrow$ (IF) $\{0 \leq i_4 \leq 1, 3 \leq i_5 \leq 3, 24 \leq i_6 \leq 40,$
                        $16 \leq i_7 \leq 24\}$

FOUR $\rightarrow$ (OF) $\{24 \le o_6 \le 56, 16 \le o_7 \le 24\}$
$\qquad$ (NF)(IF) $\{0 \le i_4 \le 1, 6 \le i_5 \le 7, 32 \le i_6 \le 48,$
$\qquad\qquad$ $08 \le i_7 \le 24\}$
FOUR $\rightarrow$ (IF) $\{1 \le i_4 \le 2, 2 \le i_5 \le 3, 40 \le i_6 \le 56,$
$\qquad\qquad$ $16 \le i_7 \le 24\},$
$\qquad$ (NF)(OF) $\{32 \le o_6 \le 48, 0 \le o_7 \le 16\}$
FOUR $\rightarrow$ (VF) $\{11 \le v_4 \le 22, 03 \le v_5 \le 03, 40 \le v_6 \le 56,$
$\qquad\qquad$ $16 \le v_7 \le 24\}$
$\qquad$ (NF)(OF) $\{32 \le o_6 \le 48, 0 \le o_7 \le 16\}$
FIVE $\rightarrow$ (VF) $\{41 \le v_4 \le 42, 44 \le v_5 \le 44, 8 \le v_6 \le 32,$
$\qquad\qquad$ $24 \le v_7 \le 30\}$
$\qquad$ (NF)(OF) $\{32 \le o_6 \le 40, 16 \le o_7 \le 24\}$
FIVE $\rightarrow$ (ZF) $\{411 \le z_4 \le 622, 404 \le z_5 \le 444, 8 \le z_6 \le 32,$
$\qquad\qquad$ $24 \le z_7 \le 30\}$
$\qquad$ (NF)(OF) $\{32 \le o_6 \le 40, 16 \le o_7 \le 24\}$
FIVE $\rightarrow$ (ZF) $\{411 \le z_4 \le 622, 404 \le z_5 \le 444, 8 \le z_6 \le 32,$
$\qquad\qquad$ $24 \le z_7 \le 30\}$
$\qquad$ (NF)(IF) $\{0 \le i_4 \le 0, 0 \le i_5 \le 0, 32 \le i_6 \le 40,$
$\qquad\qquad$ $16 \le i_7 \le 24\}$
SIX $\rightarrow$ (VF) $\{11 \le v_4 \le 42, 42 \le v_5 \le 53, 16 \le v_6 \le 32,$
$\qquad\qquad$ $24 \le v_7 \le 30\}$
SIX $\rightarrow$ (ZF) $\{111 \le z_4 \le 422, 423 \le z_5 \le 533, 16 \le z_6 \le 32,$
$\qquad\qquad$ $24 \le z_7 \le 30\}$
SEVEN $\rightarrow$ (IF) $\{3 \le i_4 \le 7, 4 \le i_5 \le 5, 32 \le i_6 \le 40, 16 \le i_7 \le 24\}$
$\qquad$ (NF)(OF) $\{32 \le o_6 \le 40, 16 \le o_7 \le 24\}$
SEVEN $\rightarrow$ (VF) $\{12 \le v_4 \le 33, 43 \le v_5 \le 43, 0 \le v_6 \le 16,$
$\qquad\qquad$ $16 \le v_7 \le 30\}$
$\qquad$ (NF)(OF) $\{32 \le o_6 \le 40, 16 \le o_7 \le 24\}$
EIGHT $\rightarrow$ (OF) $\{32 \le o_6 \le 48, 0 \le o_7 \le 16\}$
$\qquad$ (NF)(OF) $\{32 \le o_6 \le 48, 0 \le o_7 \le 16\}$
EIGHT $\rightarrow$ (IF) $\{0 \le i_4 \le 1, 2 \le i_5 \le 3, 32 \le i_6 \le 48, 0 \le i_7 \le 16\}$
$\qquad$ (NF)(OF) $\{32 \le o_6 \le 48, 0 \le o_7 \le 16\}$
NINE $\rightarrow$ (ZF) $\{321 \le z_4 \le 543, 720 \le z_5 \le 730, 32 \le z_6 \le 40,$
$\qquad\qquad$ $16 \le z_7 \le 24\}$
NINE $\rightarrow$ (VF) $\{52 \le v_4 \le 54, 74 \le v_5 \le 74, 32 \le v_6 \le 40,$
$\qquad\qquad$ $0 \le v_7 \le 16\}$
NINE $\rightarrow$ (VF) $\{21 \le v_4 \le 42, 20 \le v_5 \le 30, 32 \le v_6 \le 40,$
$\qquad\qquad$ $16 \le v_7 \le 24\}$

Recognition: A bottom-up parsing, implemented in terms of two pushdown transducers, is used.

References

1. R. De Mori, A descriptive technique for automatic speech recognition. *IEEE Trans. Audio Electroacoustics* **AU-21**, 89–100 (1972).
2. R. Newman, K. S. Fu, and K. P. Li, A syntactic approach to the recognition of liquids and glides. *Proc. Conf. Speech Commun. Process., Newton, Massachusetts, April 24–26, 1972.*
3. W. A. Lea, An approach to syntactic recognition without phonemics. *Proc. Conf. Speech Commun. Process., Newton, Massachusetts, April 24–26, 1972.*
4. A. Kurematsu, M. Takeda, and S. Inoue, *A Method of Pattern Recognition Using Rewriting Rules.* Tech. Note No. 81. Res. and Develop. Lab., Kokusai Denshiu Denwa Co. Ltd., Tokyo, Japan, June 1971.
5. M. R. Ito and R. W. Donaldson, Zero-crossing measurements for analysis and recognition of speech sounds. *IEEE Trans. Audio Electroacoustics* **AU-19**, 235–242 (1971).
6. G. D. Ewing and J. F. Taylor, Computer recognition of speech using zero-crossing information. *IEEE Trans. Audio Electroacoustics* **AU-17**, 37–40 (1969).

Appendix G

Plex Languages

In string language derivations, each terminal or nonterminal symbol may appear in a string with a symbol concatenated to itself at either the left or the right. Each symbol can be visualized as having two "attaching points," a left one and a right one (or a head and a tail), at which it may connect to or associate with other symbols. Based on an idea of Narasimhan [1], Feder has extended this to languages with symbols having an arbitrary number of attaching points for connecting to other symbols [2]. A symbol of N attaching points is called an N attaching-point entity (NAPE). Structures formed by interconnecting NAPEs are called "plex structures." Languages called plex languages can be formed from sets of plex structures. A grammar used for the specification of a plex language is called a plex grammar.

A plex grammar can be represented by a six-tuple, $[V_T, V_N, P, S, Q, q_0]$, where V_T is a finite nonempty set of NAPEs called the terminals; V_N is a finite nonempty set of NAPEs called the nonterminals; $V_T \cap V_N = \varnothing$; P is a finite set of productions or rewriting rules; $S \in V_N$ is a special NAPE called the initial NAPE; Q is a finite set of symbols called identifiers; $Q \cap (V_T \cup V_N) = \varnothing$; $q_0 \in Q$ is a special identifier called the null identifier.

The parallel between the terminals, nonterminals, productions, and initial NAPE and the components of a string language grammar should be evident. The symbols of Q are used to refer to the attaching points of the NAPEs. Every attaching point of every NAPE has an associated identifier. No two

attaching points of the same NAPE have the same identifier. The null identifier q_0 serves as a place marker and is not associated with any attaching
points. Attachment of NAPEs must occur via actual attaching points of the
NAPEs; "imaginary" connections using the null identifier are not permitted.
The number of identifiers required for the grammar is equal to one more than
the number of attaching points possessed by the NAPE with the greatest
number of attaching points.

A set of NAPEs is said to be connected if there exists a path through the
NAPEs of the set from any NAPE to any other NAPE in the set. When a
closed path can be traversed among a set of NAPEs, the set is said to contain
a loop.

The unrestricted rule of P is of the form

$$\psi\Gamma_\psi\Delta_\psi \rightarrow \omega\Gamma_\omega\Delta_\omega, \qquad \psi\Gamma_\psi, \ \omega\Gamma_\omega = \text{connected}$$

where ψ is called the definitum component list, ω the definition component
list, Γ_ψ the definitum joint list, Γ_ω the definition joint list, Δ_ψ the definitum
tie-point list, and Δ_ω the definition tie-point list.

The definitum and definition component lists are strings of the form
$\psi = a_1a_2, \ldots, a_i, \ldots, a_m$ and $\omega = b_1b_2, \ldots, b_j, \ldots, b_n$, where a_i and b_j are
single NAPEs called components. ψ and ω list and provide an ordering for the
groups of connected NAPEs that comprise the definitum and definition, respectively. The connection of attaching points of two or more NAPEs forms
a joint. The definitum and definition joint lists specify the way in which the
NAPEs of their respective component lists interconnect. The joint lists are
divided into fields. The fields are lists of identifiers, of the form $q_{i_1}q_{i_2}\cdots q_{i_k}$,
that specify which attaching points of which NAPEs connect at each joint.
One field is required per joint. The length of the fields for definitum and definition is given by $l(\psi)$ and $l(\omega)$, respectively, where l denotes the length of its
string argument. The entry q_i in the jth position of a field indicates that attaching point q_i of the jth component of the component list preceding Γ connects
at the joint. If the jth component is not involved at the particular joint, then
the null identifier q_0 appears in this position. Each joint list field must contain
at least two nonnull identifiers.

The substructures for the definitum and definition connect to the remainder
of the plex at a finite number of joints called tie points. The tie-point lists
give the correspondence between these external connectors for definitum and
definition. The tie-point lists are divided into fields; one field specifies each
tie point. Since the number of tie points for the definitum and definition must
be the same, the number of fields for both is the same. The correspondence
between tie points is given by the ordering of the fields: the tie point specified
by the pth field of the definitum tie-point list corresponds to the tie-point
specified tie points in the same way that the joint-list fields specify joints.

Both types of fields have the same length. Each tie-point list field must contain at least one nonnull identifier. In cases in which a joint is mentioned in both joint and tie-point lists, there is redundancy in the rule; information about only one of the components meeting at the joint need be furnished in the particular field to achieve a complete specification.

The following basic assumptions are made.

(1) A NAPE cannot connect itself. A grammar rule definition specifying such a connection is illegal.

(2) No interconnection among the components of a definitum or a definition other than that described in the joint list(s) can exist.

(3) Separate tie points of a definitum or definition cannot refer to the same joint or attaching point.

(4) Every attaching point of every NAPE in a grammar rule definitum (definition) must either connect with another NAPE in the definitum (definition) or be one of the tie points. It follows that every attaching point of every NAPE in a definitum (definition) must be referenced in at least one field of the definitum (definition).

The unrestricted grammar rule can specify the exchange of any substructure for any other substructure. Grammars composed of rules of this type are called unrestricted plex grammars and the languages defined by such grammars are called unrestricted plex languages. Languages and grammars of this type are analogous to unrestricted string languages and grammars.

A context-sensitive rule is obtained from the unrestricted rule by stipulating that $\psi \Gamma_\psi \Delta_\psi$ and $\omega \Gamma_\omega \Delta_\omega$ must be decomposable as follows:

$$\psi = A_{\psi_1}, \qquad \Gamma_\psi = \Gamma_{\psi_1} \Gamma_{A\psi_1}, \qquad \Delta_\psi = \Delta_A \Delta_{\psi_1},$$

$$\omega = X\psi_1, \qquad \Gamma_\omega = \Gamma_X \omega_{\psi_1} \Gamma_{X\psi_1}, \qquad \Delta_\omega = \Delta_X \Delta_{\psi_1}$$

where $X \neq$ null and $X\Gamma_X$ is connected. In the foregoing

Γ_{ψ_1} is a joint list containing fields of length $l(\psi_1)$ that describes the interconnection of the NAPEs listed as ψ_1;

Γ_X is a joint list containing fields of length $l(X)$ that describes the interconnection of the NAPEs listed as X;

$\Gamma_{A\psi_1}$ is a joint list containing fields of length $l(A) + l(\psi_1) = 1 + l(\psi_1)$; $\Gamma_{A\psi_1}$ gives the interconnection of A and the components of ψ_1, by listing the joints connecting A to $\psi_1 \Gamma_{\psi_1}$;

$\Gamma_{X\psi_1}$ is a joint list containing fields of length $l(X) + l(\psi_1)$ that gives the interconnection between $X\Gamma_X$ and $\psi_1 \Gamma_{\psi_1}$;

Δ_A and Δ_X are tie-point lists that give the correspondence between the attaching points of A and the tie points of $X\Gamma_X$;

Δ_{ψ_1} lists the tie points of $\psi_1 \Gamma_{\psi_1}$.

It is not necessary to state Δ_{ψ_1}; the context-sensitive rule can be written as follows:

$$A\psi_1\Gamma_{\psi_1}\Gamma_{A\psi_1}\Delta_A \to X\psi_1\Gamma_X\Gamma_{\psi_1}\Gamma_{X\psi_1}\Delta_X, \qquad X \neq \text{null}, \quad X\Gamma_X = \text{connected}$$

This rule declares that the NAPE A can be replaced by the subplex defined by $X\Gamma_X$ provided that A is embedded in the substructure specified by $\psi_1\Gamma_{\psi_1}$. Plex grammars containing only rules of this type and the languages defined by such grammars are called context sensitive. Such languages and grammars are generalized versions of context-sensitive string languages and grammars, respectively.

If $\psi_1 = \text{null}$ in the foregoing, then a context-free rule is obtained:

$$A\Delta_A \to X\Gamma_X\Delta_X, \qquad X \neq \text{null}, \quad X\Gamma_X = \text{connected}$$

This rule states that the NAPE A appearing in any context can be replaced by the subplex given by $X\Gamma_X$. A context-free plex grammar and context-free plex language are obtained by restriction to rules of this type.

String languages containing a finite number of strings and the grammars for such languages are referred to as finite. The analog of a finite string grammar is a finite plex grammar. The rules are of the form

$$S \to Z\Gamma_z, \qquad Z\Gamma_z = \text{connected}$$

The corresponding language is called a finite plex language. The $Z\Gamma_z$ are strings on an alphabet, $Q \cup V_T$, and can be considered to be encoded versions of the plexes that comprise the finite plex language.

Many properties of string grammars can also be extended to include plex grammars [2]. Special cases of plex grammars are obtained by imposing restrictions as follows:

(1) Terminal NAPEs can have at most two attaching points. In this case the terminal structures formed are directed graphs with labeled branches.

(2) Terminal and nonterminal NAPEs can have at most two attaching points. The PDL used by Shaw [4] is closely related to a context-free plex grammar of this type.

(3) Restriction (2) plus the restrictions that only two NAPEs can meet at a joint and that closed loops of NAPEs are prohibited in grammar rule definitions. In this case the plex grammar degenerates into an ordinary string language grammar.

The following examples are given to illustrate the use of (context-free) plex languages for pattern description. In all examples positive integers are used as identifiers. Zero is reserved as the null identifier q_0. Parentheses are used to enclose the fields comprising Γ, and Δ are separated by commas.

Example G.1 Consider a context-free plex grammar to generate patterns for the letters A and H.

(1) $\qquad$ $\langle LETTER \rangle \rightarrow \langle A \rangle$, $\langle LETTER \rangle \rightarrow \langle H \rangle$

(2) $\qquad$ $\langle A \rangle \rightarrow \langle SIDE \rangle \langle SIDE \rangle \langle ST \rangle (110, 201, 022)$

(3) $\qquad$ $\langle H \rangle \rightarrow \langle SIDE \rangle \langle SIDE \rangle \langle ST \rangle (201, 022)$

(4) $\qquad$ $\langle SIDE \rangle (1, 2) \rightarrow \langle ST \rangle \langle ST \rangle (12)(20, 12)$

The single terminal NAPE in this grammar is $\langle ST \rangle$. This NAPE represents a straight line, defined without regard to orientation, with attaching points, (denoted by the identifiers 1 and 2) at each end. The nonterminal NAPEs for the grammar are $\langle LETTER \rangle$, $\langle A \rangle$, $\langle H \rangle$, and $\langle SIDE \rangle$. Only the last of these has associated attaching points. $\langle LETTER \rangle$ is the initial NAPE.

The first rule of the grammar above defines $\langle LETTER \rangle$ as either an $\langle A \rangle$ or an $\langle H \rangle$. Rule 2 describes an A as being constructed of two $\langle SIDE \rangle$'s plus a crossbar ($\langle ST \rangle$). Three joints are formed; thus there are three fields in Γ. The first field, 110, describes the joint formed by the connection of the two $\langle SIDE \rangle$'s. This field states that attaching point 1 of the first component of the definition ($\langle SIDE \rangle$) connects to attaching point 1 of the second component of the definition ($\langle SIDE \rangle$). The zero in the third position indicates that the third component of the definition $\langle (ST) \rangle$ is not involved at the joint. The remaining two joints occur where the "sides" of the A connect to the crossbar and are described by the second and third fields. The construction of an H is described in rule 3 and is similar to that of an A except for the absence of the joint connecting the two sides. The component interconnection that takes place in rules 2–4 is illustrated in Fig. G.1.

Rule 4 of the grammar is the only one containing entries for Δ. This rule describes the joining of two $\langle ST \rangle$ components to form the nonterminal NAPE $\langle SIDE \rangle$. The single joint formed is described by $\Gamma = (12)$. Since $\langle SIDE \rangle$ has two attaching points, two tie-point list fields are needed. Attaching point 1 of $\langle SIDE \rangle$ corresponds to the tie point described by 20 (attaching point 2 of the first $\langle ST \rangle$ component). Attaching point 2 of $\langle SIDE \rangle$ corresponds to the tie point described by 12. This is the joint that is described by Γ.

Example G.2 (Grammar for describing the repetitive chemical structure of a natural rubber molecule). The structure of a natural rubber molecule is shown in Fig. G.2a. The terminal NAPEs of the grammar are shown in Fig. G.2b and consist of the carbon atom $\langle C \rangle$, with four attaching points, and the hydrogen atom $\langle H \rangle$, with one attaching point. The grammar rules are given in Fig. G.2c. The nonterminal NAPEs are $\langle CHAIN \rangle$, $\langle SECTION \rangle$, $\langle CH_3 \rangle \langle CH_2 \rangle$, and $\langle CH \rangle$. $\langle CHAIN \rangle$ is the initial NAPE. Recursion is used

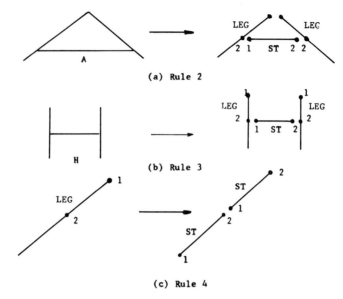

Fig. G.1. Component interconnection in A–H grammar.

in the first rule of the grammar to specify ⟨CHAIN⟩ as an arbitrary number of repetitions of ⟨SECTION⟩.

Example G.3   Figure G.3 gives a grammar that describes a simple shift register formed by assembling a number of identical shift stages. A diagram of a two-stage shift register of this type is shown in Fig. G.3a. The terminal NAPEs of the grammar are shown in Fig. G.3b and the grammar rules in Fig. G.3c. Only two productions are used. The first is recursive and describes the initial NAPE, ⟨SHIFT RGSTR⟩, as consisting of any number of repetitions of the nonterminal ⟨SHFT STGE⟩. The second rule describes ⟨SHFT STGE⟩ in terms of terminal NAPEs.

Example G.4   Figure G.4 gives two grammars for generating flow-charts composed of start, halt, function, and predicative (decision) blocks. The terminal NAPEs are shown in Fig. G.4a and the grammar rules for generating flow charts without and with loops are shown in Fig. G.4b and Fig. G.4c, respectively.

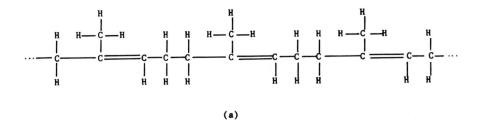

(a)

(b)

(1) $\langle CHAIN \rangle (1, 2) \rightarrow \langle SECTION \rangle \langle CHAIN \rangle (21)(10, 02)$
$\langle CHAIN \rangle (1, 2) \rightarrow \langle SECTION \rangle ()(1, 2)$
(2) $\langle SECTION \rangle (1, 2) \rightarrow \langle CH_2 \rangle \langle C \rangle \langle CH_3 \rangle \langle CH \rangle \langle CH_2 \rangle (21000,$
$02100, 03020, 04010, 00031)(10000, 00002)$
(3) $\langle CH_3 \rangle (1) \rightarrow \langle CH_2 \rangle \langle H \rangle (21)(10)$
(4) $\langle CH_2 \rangle (1, 2) \rightarrow \langle CH \rangle \langle H \rangle (21)(10, 30)$
(5) $\langle CH \rangle (1, 2, 3) \rightarrow \langle C \rangle \langle H \rangle (41)(10, 20, 30)$

(c)

Fig. G.2. Grammar for natural rubber Molecule. (a) Natural rubber molecule; (b) terminal NAPEs; (c) grammar rules.

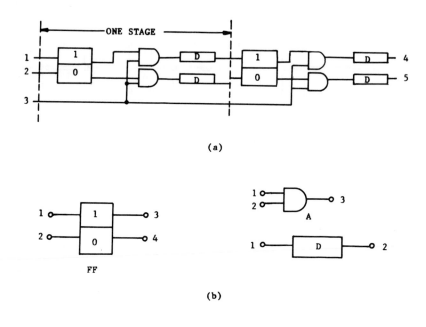

(a)

(b)

(1)   ⟨SHFT RGSTR⟩(1, 2, 3, 4, 5) → ⟨SHFT STGE⟩⟨SHFT RGSTR⟩
                                                (41, 52, 33)(10, 20, 33, 04, 05)
        ⟨SHFT RGSTR⟩(1, 2, 3, 4, 5) → ⟨SHFT STGE⟩()(1, 2, 3, 4, 5)
(2)   ⟨SHFT STGE⟩(1, 2, 3, 4, 5) → ⟨FF⟩⟨A⟩⟨A⟩⟨D⟩⟨D⟩
                                                (31000, 40100, 02200, 03010, 00301)
                                                (10000, 20000, 02200, 00020, 00002)

(c)

Fig. G.3. Grammar for simple shift register (a) Simple shift register; (b) terminal
NAPEs; (c) grammar rules.

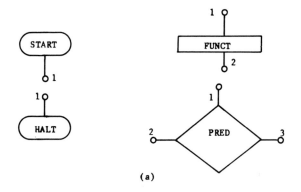

(a)

(1) ⟨PROG⟩ → ⟨START⟩⟨END⟩(11)
(2) ⟨END⟩(1) → ⟨HALT⟩(1), ⟨END⟩(1) → ⟨FUNCT⟩⟨END⟩(21)(10)
     ⟨END⟩(1) → ⟨PRED⟩⟨END⟩⟨END⟩(210, 201)(100)

(b)

• (1) ⟨PROG⟩ → ⟨START⟩⟨P⟩⟨HALT⟩(110, 021)
  (2) ⟨P⟩(1, 2) → ⟨FUNCT⟩()(1, 2), ⟨P⟩(1, 2) → ⟨FUNCT⟩⟨P⟩(21)(10, 02),
       ⟨P⟩(1, 2) → ⟨PRED⟩⟨P⟩(21, 12)(21, 30),
       ⟨P⟩(1, 2) → ⟨PRED⟩⟨P⟩⟨P⟩(210, 301, 022)(100, 022)

(c)

Fig. G.4. Flowchart grammars. (a) Terminal NAPEs; (b) grammar for flow charts containing no loops; (c) grammar for flow charts containing loops.

References

1. R. N. Narasimhan, Syntax-directed interpretation of classes of pictures. *Comm. ACM* **9**, 166–173 (1966).
2. J. Feder, Plex languages. *Information Sci.* **3**, 225–241 (1971).
3. J. Feder, Languages of encoded line patterns. *Information and Control* **13**, 230–244 (1968).
4. A. C. Shaw, A formal picture description scheme as a basis for picture processing system. *Information and Control* **14**, 9–52 (1969).

Appendix H

Web Grammars

One of the two-dimensional grammars is the web grammar proposed by Pfaltz and Rosenfeld [1]. Sentences generated by a web grammar are directed graphs with symbols at their vertices ("webs"). A web grammar G is a four-tuple

$$G = (V_N, V_T, P, S)$$

where V_N is a set of nonterminals, V_T a set of terminals, S a set of "initial" webs, and P a set of web productions or rewriting rules. A web production is defined as[†]

$$\alpha \to \beta, E$$

where α and β are webs, and E is an embedding of β. If we want to replace the subweb α of the web ω by another subweb β, it is necessary to specify how to "embed" β in ω in place of α. The definition of an embedding must not depend on the host web ω, since we want to be able to replace α by β in any web containing α as a subweb. Usually E consists of a set of logical functions which specify whether or not each vertex of $\omega - \alpha$ is connected to each vertex of β.

[†] In a most general formulation, the contextual condition of the production is added [1, 2].

Example H.1 Consider a web grammar $G = (V_N, V_T, P, S)$, where $V_N = \{A\}$, $V_T = \{a, b, c\}$, $S = \{\dot{A}\}$, and P:

(1) $\dot{A} \longrightarrow$

$E = \{(p, a) \,|\, (p, A)$ an edge in the host web$\}$

(2) $\dot{A} \longrightarrow$

E is the same as in (1)

The language of this grammar is the set of all webs of the form

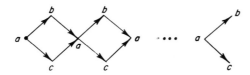

It is noted that web grammars are vertex or node oriented compared with the branch- or edge-oriented grammars (e.g., PDL, plex grammars). That is, terminals or primitives are represented as vertices in the graph rather than as branches.

An important special case of a web grammar is that in which the terminal set V_T consists of only a single symbol. In this case, every point of every web in the language has the same label, so that we can ignore the labels and identify the webs with their underlying graphs. This type of web grammar is called a "graph grammar," and its language is called a graph language. A web production is context sensitive if there exists a point a of α such that $\alpha - \{a\}$ is a subweb of β and all edges between points of the host web and points of $\alpha - \{a\}$ are in E. In particular, the production will be context free if α has only a single point. Thus, a web grammar is called context sensitive (context free) if all its productions are context sensitive (context free). The web grammar in Example H.1 is context free since only one-point webs are rewritten.

Comparing web grammars with the plex grammars, we can consider the NAPEs in a plex grammar as webs in which one point is labeled with the name of the NAPE and the others with the identifiers of its attaching points. The joint lists in a plex grammar, which describe how sets of NAPEs are interconnected, correspond to the edges internal to the subwebs α and β in a web production, while the tie-point list corresponds to the embedding E of β in the host web.

Example H.2   Consider the context-free graph grammar

$$G = (V_N, V_T, P, S)$$

were

$$V_N = \{A\}, \quad V_T = \{a\}, \quad S = \{a, \overset{\bullet \longrightarrow}{a \quad A}\}$$

and P:

(1) $\overset{\bullet \longrightarrow \quad \longrightarrow}{a \quad a \quad A}$ $E = \{(p, a) \mid (p, A)$ an edge in the host web$\}$

(2) $\overset{\bullet \longrightarrow \bullet}{A \quad A}$
    $ A$ $E = \{(p, A) \mid (p, A)$ an edge in the host web$\}$

(3) $\overset{\bullet \longrightarrow \bullet}{A \quad a}$ $E = \{(p, a) \mid (p, A)$ an edge in the host web$\}$

The language generated by this web grammar consists of all directed trees which have least elements.

Example H.3   The following context-free graph grammar generates the set of webs that includes the class of all basic two-terminal series–parallel networks (TTSPN)†:

$$G = (V_N, V_T, P, S)$$

where

$$V_N = \{A\}, \quad V_T = \{a\}, \quad S = \{\overset{\longrightarrow}{a \quad a}, \overset{\longrightarrow \quad \longrightarrow}{a \quad A \quad a}\}$$

and P:

(1) $\overset{\bullet \longrightarrow \quad \longrightarrow}{A \quad A_{(1)} \quad A_{(2)}}$ $E = \{(p, A_{(1)}) \mid (p, A)$ an edge in the host web$\} \cup \{(A_{(2)}, p) \mid (A, p)$ an edge in the host web$\}$

(2) $\overset{\bullet \longrightarrow \bullet}{A \quad A}$
    $ A$ $E = \{(p, A) \mid (p, A)$ an edge in the host web$\} \cup \{(A, p) \mid (A, p)$ an edge in the host web$\}$

(3) $\overset{\bullet \longrightarrow \bullet}{A \quad a}$ E is the same as in (2)

† It should be noticed that the grammar will generate only the set of all basic TTSPN's if we impose a condition on production (2) that the left and right contexts of A both have cardinality one. Of course, with this modification, the grammar becomes context sensitive [3].

A typical TTSPN generated, for example, would be

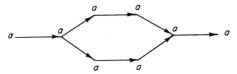

Example H.4 Consider the context-sensitive web grammar $G = (V_N, V_T, P, S)$ where $V_N = \{B, Z, Z', W, W', X\}$, $V_T = \{a, b, c, w, x, y, z\}$,

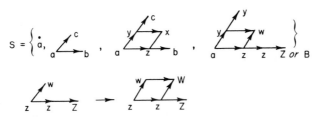

where the embedding of the "new" vertex, labeled W, is as shown; the attachments or connections of the other vertices in the rewritten web remain unchanged. Similar embeddings occur in the remaining rules.

The grammar generates "directed triangles" of the form

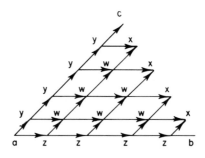

The syntax analysis (parsing) for the grammar generating all TTSPNs was implemented using the language GIRL (Graph Information Retrieval Language [4]. GIRL is a SLIP-like extension of FORTRAN which allows a programmer to store a directed graph data structure in the computer by using a few basic instructions. The analysis of a set of black-and-white images simulating neural nets in terms of a web grammar was implemented, using a system of FORTRAN programs collectively called NRNTST (for "neuron test"). NRNTST would accept pictures of simulated neural nets and produce corresponding picture description webs [4].

References

1. J. L. Pfaltz and A. Rosenfeld, WEB grammars. *Proc. Int. Joint Conf. Artificial Intelligence 1st, Washington, D.C., May 1969*, pp. 609–619.
2. G. U. Montanari, *Separable Graphs, Planar Graphs and Web Grammars*. Tech. Rep. 69–96. Comput. Sci. Center, Univ. of Maryland, College Park, 1969.
3. J. L. Pfaltz and A. Rosenfeld, Private communication, 1973.
4. J. L. Pfaltz, *Web Grammars and Picture Description*. Tech. Rep. 70–138. Comput. Sci. Center, Univ. of Maryland, College Park, 1970.

Appendix I

Tree Grammars for Syntactic Pattern Recognition

This appendix presents a brief introduction to tree grammars and tree automata and their application to syntactic pattern recognition.

Definition I.1 Let N^+ be the set of strictly positive integers. Let U be the universal tree domain (the free semigroup with identity element "0" generated by N^+ and a binary operation "·"). Figure I.1 represents the universal tree domain U.

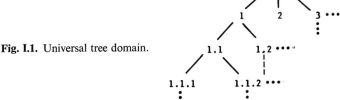

Fig. I.1. Universal tree domain.

Definition I.2 A ranked alphabet is a pair $\langle \Sigma, r \rangle$ where Σ is a finite set of symbols and

$$r: \quad \Sigma \rightarrow N = N^+ \cup \{0\}$$

For $a \in \Sigma$, $r(a)$ is called the rank of a. Let $\Sigma_N = r^{-1}(n)$.

Definition I.3   A tree over Σ (i.e., over $\langle \Sigma, r \rangle$) is a function

$$\alpha: \quad D \to \Sigma$$

such that D is a tree domain and

$$r[\alpha(a)] = \max \{i \,|\, a \cdot i \in D\}$$

The domain of a tree α is denoted by $D(\alpha)$. Let T_Σ be the set of all trees over Σ.

Definition I.4   Let α be a tree and a be a member of $D(\alpha)$. α/a, a subtree of α at a, is defined as

$$\alpha/a = \{(b, x) \,|\, (a \cdot b, x) \in \alpha\}$$

Definition I.5   A regular tree grammar over $\langle V_T, r \rangle$ is a four-tuple

$$G_t = (V, r', P, S)$$

satisfying the following conditions:

  (i) $\langle V, r' \rangle$ is a finite ranked alphabet with $V_T \subseteq V$ and $r'/V_T = r$. $V - V_T = V_N$, the set of nonterminals.
 (ii) P is a finite set of productions of the form $\Phi \to \psi$ where Φ and ψ are trees over $\langle V, r' \rangle$.
(iii) S is a finite subset of T_V, where T_V is the set of trees over alphabet V.

Definition I.6 $\alpha \overset{a}{\Rightarrow} \beta$ is in G_t if and only if there exists a production $\Phi \to \psi$ in P such that Φ is a subtree of α at a and β is obtained by replacing the occurrence of Φ at a by ψ. We write $\alpha \Rightarrow \beta$ in G_t if and only if there exists $a \in D(\alpha)$ such that $\alpha \overset{a}{\Rightarrow} \beta$.

Definition I.7 $\alpha \overset{*}{\Rightarrow} \beta$ is in G_t if and only if there exists $\alpha_0, \alpha_1, \ldots, \alpha_m$, $m > 0$, such that

$$\alpha = \alpha_0 \Rightarrow \alpha_1 \Rightarrow \cdots \Rightarrow \alpha_m = \beta$$

in G_t. The sequence $\alpha_0, \ldots, \alpha_m$ is called a derivative or deduction of β from α, and m is the length of the deduction.

Definition I.8 $L(G_t) = \{\alpha \in T_{V_T} | $ there exists $Y \in S$ such that $Y \overset{*}{\Rightarrow} \alpha$ in $G_t\}$ is called the (tree) language generated by G_t.

Definition I.9   A tree grammar $G_t = (V, r, P, S)$ is expansive if and only if

each production in P is of the form

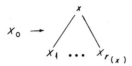

where $x \in V_T$ and $X_0, X_1, \ldots, X_{r(x)}$ are nonterminal symbols.

Theorem I.1 For each regular tree grammar G_t, one can effectively construct an equivalent expansive grammar G_t'; that is, $L(G_t') = L(G_t)$ [1].

Definition I.10 A tree automaton over $\sum$ is a $(k + 2)$-tuple

$$M_t = (Q, f_1, \ldots, f_k, F)$$

where (i) Q is a finite set of states; (ii) for each i, $1 \le i \le k$, f_i is a relation on $Q^{r(\sigma_i)} \times Q$, $\sigma_i \in \Sigma$, that is $f_i\colon Q^{r(\sigma_i)} \to Q$; and (iii) $F \subseteq Q$ is a set of final states.

Definition I.11 The response relation ρ of a tree automaton M_t is defined as

(i) if $\sigma \in \Sigma_0$, $\rho(\sigma) \sim X$ if and only if $f_\sigma \sim X$, that is, $\rho(\sigma) = f_\sigma$;
(ii) if $\sigma \in \Sigma_n$, $n > 0$, $\rho(\sigma, x_0, \ldots, x_{n-1}) \sim X$ if and only if there exists $x_0, \ldots, x_{n-1}$ such that $f_\sigma(x_0, \ldots, x_{n-1}) \sim X$ and $\rho(x_i) \sim X_i$, $1 \le i \le n$, that is, $\rho(\sigma, x_0, \ldots, x_{n-1}) = f_\sigma(\rho(x_{n-1}))$.

Definition I.12 $T(M_t) = \{\alpha \in T_\Sigma |$ there exists $X \in F$ such that $\rho(\alpha) \sim X\}$ is called the set of trees accepted by M_t.

Theorem I.2 For every regular tree grammar G_t, one can effectively construct a tree automaton M_t such that $T(M_t) = L(G_t)$ [1].

The construction procedure is summarized as follows:

(i) Obtain an expansive tree grammar $G_t' = (V', r, P'S)$ for the given regular tree grammar $G_t = (V, r, P, S)$ over alphabet V_T.
(ii) The equivalent (nondeterministic) tree automaton is

$$M_t = (V' - V_T, f_1 \ldots, f_k, \{S\})$$

where $f_x(X_1, \ldots, X_n) \sim X_0$ if $X_0 \to xX_1, \ldots, X_n$ is in P'. The application of tree grammars and tree automata to the description and recognition of patterns is illustrated by the following examples [2].

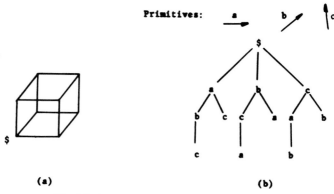

Fig. I.2. Tree representation of a square object.

Example I.1   The square object in Fig. I.2a can be described by the tree shown in Fig. I.2b.

Example I.2   The tree grammar $G_t = (V, r, P, S)$, where $V = \{S, a, b, \$, A, B\}$, $V_T = \{\xrightarrow{a}, \uparrow b, \cdot \$\}$, $r(a) = \{2, 1, 0\}$, $r(b) = \{2, 1, 0\}$, $r(\$) = 2$, and P:

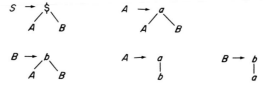

generates such patterns as

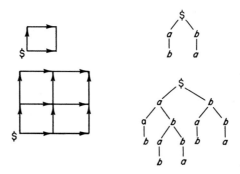

The tree automaton which accepts the set of trees generated by G_t is

$$M_t = (Q, f_a, f_b, f_\$, F)$$

where $Q = \{q_a, q_b, q, q_F\}$, $F = \{q_F\}$, and f:

$$f_a = q_a \quad f_a(q,q) = q \quad f_a(q_b) = q$$
$$f_b = q_b \quad f_b(q,q) = q \quad f_b(q_a) = q$$
$$f_\$(q,q) = q_F$$

Example I.3 The following tree grammar can be used to generate trees representing the *L–C* networks shown in Fig. I.3.

$$G_t = (V, r, P, S)$$

where

$$V_T = \{S, \; \odot \; V_{in}, \; \text{-}\widehat{\text{mm}}\text{-} \; L, \; \dashv\vdash \; C, \; \dashv\!\!\! \cdot \; W, \cdot \; \$\}$$
$$r(V_{in}) = 1, \quad r(L) = \{2, 1\}, \quad r(C) = 1, \quad r(W) = 0, \quad r(\$) = 2$$

and *P*:

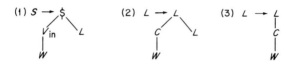

For example, after applying productions (1), (2), and (3), the following tree is generated.

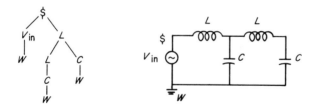

The tree automaton which accepts the set of trees generated by G_t is

$$M_t = (Q, f_{V_{in}}, f_L, f_C, f_W, f_\$, F)$$

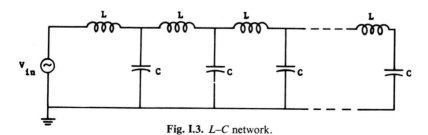

Fig. I.3. *L–C* network.

where $Q = \{q_1, q_2, q_3, q_4, q_F\}$, $F = \{q_F\}$, and f:

$$f_{V_{1n}}(q_1) = q_4, \qquad f_L(q_3) = q_2$$
$$f_L = q_2, \qquad f_L(q_2, q_3) = q_2$$
$$f_C(q_1) = q_3$$
$$f_W = q_1$$
$$f_\$(q_2, q_4) = q_F$$

References

1. W. S. Brainerd, Tree generating regular systems." *Information and Control* **14**, 217–231 (1969).
2. K. S. Fu and B. K. Bhargava, Tree systems for syntactic pattern recognition. *IEEE Trans. Comput.* **C22** 1087–1099 (1973).
3. J. E. Doner, Tree acceptors and some of their applications. *J. Comput. System Sci.* **4** (1970).
4. J. W. Thatcher and J. B. Wright, Generalized finite automata theory with an application to a decision problem of second order logic. *J. Math. System Theory* **2** (1969).
5. W. C. Rounds, Context free grammar on trees. *IEEE Annu. Symp. Switching and Automata Theory, 10th, October 1969*, pp. 143–148.
6. W. C. Rounds, Mappings and grammars on trees. *J. Math. System Theory* **4**, 257–287 (1970).

Author Index

A

Aho, A. V,. 25, *46*, *123*, 142(23), 145(23), *164*
Aiserman, M. A., 7(18), *21*, *23*
Ambler, A. P., 92(19), *122*
Amoss, J. O., *89*
Anderson, R. H., 19(40), *22*, 79(64), *89*, 118(42), *122*, *229*, 245, *252*
Anderson, T. W., 179(7), *192*
Andrews, H. C., 1(8), 7(8,23), *21*
Arkadev, A. G., 7(20), *21*

B

Bach, E., 53(14), *86*
Bakis, R., 16(32), *22*

Banerji, R. B., *24*, 91(8), *121*, 124(1), *163*
Bar-Hillel, Y., 25(5,6), *46*
Barrow, H. G., *24*, 68(55), *88*, 92(19), *122*
Becker, P. W., *24*
Berstein, M. I., 247(4), *252*
Bhargava, B. K., 285(2), *288*
Biermann, A. W., 194(33,34,35,36,) 198(35), 205, 212, *228*
Binford, T. O., *23*
Bobrow, D. G., 19(46), *22*
Bongard, M. M., *23*
Booth, T. L., *46*, 128(10), *164*, *165*, 227(45), 229
Brainerd, W. S., 78(75), *89*, 285(1), *288*
Braverman, E. M., 7(18,20), *21*, *23*
Brice, C. R., 63, *88*
Bruce, G. D., *164*
Burstall, R. M., 92(19), *122*
Butler, J. W., 57(27), *87*
Butler, M. K., 57(27), *87*

289

Subject Index